Camiel H.C. Janssen, Léon P.B.M. Janssen, Marijn M.C.G. Warmoeskerke.
Transport Phenomena Data Companion

Also of Interest

Chemical Kinetic Data.
Russo, Tesser, Di Serio, 2024
ISBN 978-3-11-046428-3, e-ISBN 978-3-11-046430-6

Mass, Momentum and Energy Transport Phenomena.
A Consistent Balances Approach
Van den Akker, Mudde, 2023
ISBN 978-3-11-124623-9, e-ISBN 978-3-11-124657-4

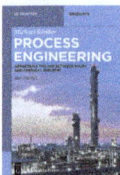

Process Engineering.
Addressing the Gap between Study and Chemical Industry
Kleiber, 2023
ISBN 978-3-11-102811-8, e-ISBN 978-3-11-102814-9

Process Technology.
An Introduction
De Haan, Padding, 2022
ISBN 978-3-11-071243-8, e-ISBN 978-3-11-071244-5

Chemical Reaction Engineering.
A Computer-Aided Approach
Salmi, Wärnå, Hernández Carucci, de Araújo Filho, 2023
ISBN 978-3-11-079797-8, e-ISBN 978-3-11-079798-5

Camiel H.C. Janssen, Léon P.B.M. Janssen,
Marijn M.C.G. Warmoeskerken[†]

Transport Phenomena Data Companion

Balances, Fluid Flow, Heat and Mass Transfer, Materials
Properties

2nd edition

DE GRUYTER

Authors
Camiel H.C. Janssen
Associate Professor
Department of Chemical Engineering
Faculty of Chemistry
National Autonomous University of
Mexico (UNAM)
Mexico

Léon P.B.M. Janssen
Professor emeritus
Department of Chemical Engineering
Faculty of Science and Engineering
University of Groningen
The Netherlands

Marijn M.C.G. Warmoeskerken†
Professor obiit
Engineering of Fibrous Smart Materials
Faculty for Engineering Technology
University of Twente
The Netherlands

About the cover picture:
The cover picture originates from "De Re Metallica", a series of twelve books about the mining and treatment of metal ores. The books were written by Georgius Agricola (Georg Bauer) and in 1556 published by 'Frobenius et Episcopus' (Froben und Bischoff) in Basil, Switzerland. The books describe the mining and purification of metals and many engineering parts have a close relationship with the present science of 'Transport Phenomena'. The cover drawing shows a grinder (K) followed by a cascade of stirred tanks in series (O) in which the ore is separated from the other materials. Different types of stirrers (S) can be used. The stirring power originates from the water wheel (B) and different flow meters (Z) that are shown.

Trivia: The books of Agricola were for the first time translated from Latin to English in 1912 by the mining engineer Herbert Hoover and his wife the geologist Lou Henry; the couple that later became the president and first lady of the United States.

ISBN 978-3-11-138507-5
e-ISBN (PDF) 978-3-11-138534-1
e-ISBN (EPUB) 978-3-11-138567-9

Library of Congress Control Number: 2024933644

Bibliographic information published by the Deutsche Nationalbibliothek
The Deutsche Nationalbibliothek lists this publication in the Deutsche Nationalbibliografie; detailed bibliographic data are available on the Internet at http://dnb.dnb.de.

© 2024 Walter de Gruyter GmbH, Berlin/Boston
Cover image: akg-images/De Agostini/Icas94
Typesetting: Integra Software Services Pvt. Ltd.

www.degruyter.com

Preface to the first edition

Many of the data needed for calculations in chemical engineering are dispersed over the literature. Moreover, various systems of units are used, often forcing the user to perform tedious conversions before a 'quick' calculation can be started. In this companion the authors have compiled those data that, according to their experience, are used frequently in transport phenomena and related subjects and, since all data are in S.I. units, rapid access to various calculations is facilitated.

This book is of course no substitute for a complete literature survey and no compendium on all data available in literature. On many occasions a selection had to be made from a multitude of expressions. For instance, for the mass transfer to bubbles, drops and particles, dozens of correlations are available. In those cases the most general correlation or the expression most commonly used has been chosen.

This companion consists of four parts. The first part is general and gives information varying from the Greek alphabet to calibration curves for thermocouples and pH ranges of indicators.

The second part consists of frequently used mathematics. In addition to general mathematical techniques, a selection of vectorial and tensorial calculus relevant to hydrodynamics and elementary rheology has been added.

The third part is a compendium of the transport phenomena. A systematic arrangement facilitates its use. The figures are in such a form that easy reading and accuracy are combined.

In the final part various material properties are given. Special attention has been paid to the most commonly used materials: air and water, but also frequently used materials like for instance hydrocarbons, foods and others are included.

For easy access to the data an extensive index is very important, so, special attention has been given to make the index as complete as possible.

We hope this book will be useful for all those involved in transport phenomena, students, scientists as well as engineers, and we are grateful to everybody who has contributed to improve this companion by suggestions and criticism. Any further suggestions and amendments will be gratefully received.

Groningen and Delft,
L.P.B.M. Janssen
M.M.C.G. Warmoeskerken

https://doi.org/10.1515/9783111385341-202

Preface to the revised edition

This Data Companion appeared as a simple compendium in 1982 in the Dutch language for students at Delft University of Technology. In 1987 this work was completely revised, expanded and translated into English for a co-edition between 'Delft University Press' and 'Edward Arnold Publishers Ltd London'. Although during the following years physical constants and material properties of course remained the same, the emphasis in Transport Phenomena did change. New areas of interest have emerged and some of the old contents of the Data Companion were hardly used anymore. Furthermore, after about forty years, the data presented needed to be brought up to a more modern format and therefore the book has been modified and rewritten.

In this new edition parts of the previous version have been omitted (like the properties of thermojunctions and advanced modifications of the Bessel functions). New parts have been added to this revised edition: amongst others, we added as a first chapter a quick reference list and added the tools for fast and approximate calculations in transport phenomena. Also more emphasis is given to frequently used differential equations; an extra chapter is added to illustrate the connections between elementary engineering problems and their associated differential equations and solutions. The list of dimensionless correlations has been rearranged and extended.

The last decades have seen an increased interest in energy and environment. Therefore we added data that are useful in solving transport phenomena problems related to CO_2 emission, green house effects and in the application of solar energy. For easier handling and quick reference at various places separate symbol lists have been added to individual chapters.

December 2023

Mexico-City and Groningen
Camiel H.C. Janssen
Léon P.B.M. Janssen

https://doi.org/10.1515/9783111385341-203

Contents

Preface to the first edition —— V

Preface to the revised edition —— VII

Chapter 1
Quick engineering and orders of magnitude —— 1
    Physical constants at 20 ^{0}C —— 1
    Orders of magnitude —— 1
    Fluid flow —— 1
        Pressure drop in pipes —— 1
        Darcy's law for porous media —— 2
        Flow forces on an object —— 2
        Simplified Bernoulli's equation (incompressible media) —— 2
    Heat transfer —— 2
        Internal versus external heat transfer —— 2
        Stationary heat transfer —— 3
        Heat transfer coefficients: h [W/m^2 K] —— 3
        Instationary heat transfer —— 4
        Radiation —— 4
    Mass transfer —— 5
        Stationary mass transfer —— 5
        Instationary mass transfer —— 6
        List of symbols and abbreviations —— 6

Chapter 2
Miscellaneous —— 7
    Greek alphabet —— 7
    Système Internationale (International System of Units) —— 7
        SI base units —— 7
        SI named units —— 7
        SI accepted units —— 8
        SI prefixes —— 9
    Periodic table —— 10
    Mathematical constants —— 13
    Physical constants —— 13
        Values of the gas constant —— 14
    Conversion factors —— 15
        Temperature conversion —— 17
    Temperature scales —— 17
    Approximations for physical properties of water and air —— 18

Notation —— **18**
Indices —— **18**
Water —— **18**
Water vapor —— **19**
Humid air —— **19**
Dry air —— **19**
pH range of acid-base indicators —— **20**

Chapter 3
Mathematics —— 21
Quadratic and cubic relations —— **21**
Volume and surface of some bodies —— **22**
Integrals —— **25**
Basic integrals —— **25**
Functions of $a + bx$ —— **26**
Functions of $a + bx^2$ —— **27**
Functions of $a + bx + cx^2$ —— **27**
Functions of $\sqrt{a + bx}$ —— **28**
Functions of $\sqrt{a^2 + x^2}$ —— **29**
Functions of $\sqrt{a^2 - x^2}$ —— **29**
Function of $\sqrt{x^2 - a^2}$ —— **30**
Transcendental functions —— **30**
Elementary differential equations —— **33**
First order —— **33**
Second order —— **34**
Second order, general —— **35**
Approximations for Bessel's functions —— **36**
The error function —— **39**
Laplace transforms —— **41**
Notation —— **41**
Definition —— **41**
Properties —— **41**
Frequently used Laplace transforms —— **43**
Vector and tensors —— **44**
Vector mathematics —— **44**
Tensor mathematics —— **44**
Invariants of a tensor —— **45**
Other vector and tensor operations —— **46**
Differentiation in vector and tensor notation —— **47**
Notation —— **47**
Nabla operator —— **47**
Nabla operations in a Cartesian coordinate system —— **47**

Nabla identities —— **48**
Scalar derivatives with respect to time —— **49**
Tensorial derivatives with respect to time —— **49**
Linear regression —— **50**
"Curve fit" of linear, exponential, and power functions —— **50**

Chapter 4
Transport phenomena —— 53
Concentration notation —— **53**
Standard quantities —— **53**
Derived quantities —— **53**
Relations —— **53**
Microscopic balances in general form —— **54**
Bernoulli's equations —— **57**
Microscopic balances for idealized materials —— **59**
Notation —— **59**
Continuity equations (total mass balances) —— **59**
Navier-Stokes equations (momentum balances) —— **59**
Fourier equations (thermal energy balances) —— **61**
Fick's equations (component mass balances) —— **62**
Continuity equation and momentum balance for special cases —— **63**
Continuity equation for compressible flow —— **63**
Momentum balance for incompressible non-Newtonian fluids —— **63**
Macrobalances —— **66**
Unsteady-state balances —— **66**
Steady-state balances —— **66**
Elementary differential equations for flow, heat transfer, and mass transfer —— **68**
Fluid flow —— **68**
Heat transfer —— **73**
Mass transfer —— **77**
Frequently occurring flow fields in tensor notation —— **83**
Rheological models —— **84**
Notation —— **84**
One-dimensional —— **84**
Tensorial —— **84**
Pressure-throughput characteristics for laminar flow of liquids in a straight round tube —— **86**
Notation —— **86**
Residence time distribution —— **87**
Quantities and their dimensions —— **90**
Alphabetical list of dimensionless numbers —— **92**

Frequently used dimensionless correlations —— **97**

List of symbols and abbreviations —— **101**

Analogies between heat transfer and mass transfer —— **102**

Drag coefficient C_w for flow around obstacles —— **103**

Drag coefficients —— **103**

Packed and fluidized beds —— **104**

Drag coefficient in a packed bed as function of Reynolds based on the hydraulic diameter —— **105**

Friction coefficient K_w for flow through tube systems —— **107**

Friction factor for flow in tubes —— **109**

Friction factors for pressure flow in pipes —— **110**

Hydraulic diameters and Reynolds numbers —— **111**

The power number Po of impellers as a function of Re —— **112**

Humidity diagram for air-water systems at atmospheric pressure (I) —— **113**

Humidity diagram for air-water systems at atmospheric pressure (II) —— **114**

Fourier instationary heat and mass transfer —— **115**

Heat transfer —— **115**

Mass transfer —— **115**

Fourier instationary heat and mass transfer (center T or C) —— **116**

Fourier instationary heat and mass transfer (average T or C) —— **117**

Shapes of free rising bubbles or drops in Newtonian liquids —— **118**

Rising Velocity of Air Bubbles in Water —— **119**

Rising velocity or fall velocity of drops in liquids of low viscosity $\eta_c < 5$ mPa s —— **120**

Two-phase gas-liquid flow in horizontal tubes —— **121**

Two-phase cocurrent flow of gas-liquid through vertical tubes —— **123**

Formation of bubbles at a nozzle —— **124**

In stationary liquid —— **124**

In flowing liquid —— **124**

Mass transfer with first-order chemical reaction —— **125**

Symbols —— **125**

Radiation —— **126**

Stefan-Boltzmann law —— **126**

Wien's displacement law —— **126**

Kirchhoff's law —— **126**

Planck's law —— **126**

Blackbody radiation exchange —— **126**

Wavelength-temperature scale of light sources —— **127**

Color temperature scale (in kelvin) —— **127**

The electromagnetic spectrum —— **128**

Emission coefficients of various materials —— **128**

Environmental data —— **130**

General —— **130**

Atmosphere —— **131**
Solar energy —— **132**
Energy storage —— **133**
Heating values —— **134**

Chapter 5
Material properties —— **135**
Water —— **135**
Ice —— **143**
Air —— **144**
Physical properties of some materials —— **147**
Notation —— **155**
Foods —— **156**
Compressibility of gases —— **158**
Antoine's parameters —— **161**
Solubility table —— **162**
Solubility parameters —— **165**
Diffusion coefficients —— **166**
Henry's law —— **168**
Henry's law constant for the solubility of gases in water —— **168**
Henry's law constant for the solubility of gases in ionic solutions —— **169**
Equilibrium data for hydrocarbons —— **170**
Dynamic viscosity —— **175**
Surface tension —— **182**

Index —— **185**

Chapter 1
Quick engineering and orders of magnitude

Physical constants at 20 °C

	Viscosity, μ (Pa·s)	Heat conductivity, λ (W/m·K)	Density, ρ (kg/m³)	Specific heat, C_p (kJ/kg·K)	Thermal diffusivity, a (10^{-7} m²/s)	Prandtl number, Pr (–)
Air	1.8×10^{-5}	26×10^{-3}	1.2	1.0	220	0.71
Water	1.0×10^{-3}	0.60	1,000	4.2	1.4	7.0
Ice (0 °C)	–	2.1	917	2.1	11	
Ethanol	1.2×10^{-3}	0.17	790	2.4	0.88	17
Glycerol	1.5	0.29	1261	2.4	0.95	12,400
Foods	1.5–800	0.2 – 0.3	830–1,100	1–2	1–3	
Concrete		0.1–0.2	1,800–2,400	1.1	0.5–1	
Wood		0.1–0.2	500–900	1.9–2.7	0.3–2	
Polymers	100–1,000	0.15–0.3	830–1,100	1–2	1–3	
Metals		20–400	2,700–21,000	0.1–1	200–1,700	

Orders of magnitude

Thermal diffusivity for common liquids $a = 10^{-7}$ m²/s

Density of air: $\rho = 1\,\text{kg/m}^3$

Density of water: $\rho = 1{,}000\,\text{kg/m}^3$

Viscosity of air: $\mu = 20\,\mu\text{Pa} \cdot \text{s}$

Viscosity of water: $\mu = 1\,\text{mPa} \cdot \text{s}$

Fluid flow

Pressure drop in pipes

$Re = \dfrac{\rho v d}{\mu}$ if Re < 2,100 laminar flow

if Re > 4,000 turbulent flow

$\Delta p = 4f \cdot \dfrac{1}{2}\rho v^2 \dfrac{L + L_e}{d}$ Laminar flow (Re < 2,100): $4f = \dfrac{64}{Re}$ (Hagen-Poiseuille)

Turbulent flow (Re > 4,000): $4f = 0.316\,Re^{-0.25}$ (Blasius)

https://doi.org/10.1515/9783111385341-001

Darcy's law for porous media

$$\Delta p = \frac{\mu L}{k_p A} \Phi \text{ with } \mathrm{Re} = \frac{\rho v d_p}{\mu} < 10$$

Flow forces on an object

$$F = C_\mathrm{w} \, A_\perp \, \frac{1}{2}\rho v^2 \qquad C_\mathrm{w} = 0.2 \text{ (sphere at Re} > 300{,}000)$$

$$\text{up to } \; C_\mathrm{w} = 2 \quad \text{(square rod at Re} > 35{,}000)$$

Simplified Bernoulli's equation (incompressible media)

$$-\left\{\frac{\Delta p}{\rho} + g\Delta h + \frac{1}{2}\Delta\left(\langle v\rangle^2\right)\right\}\Phi_m + \Phi_A - \left(4f \cdot \sum_i \left(\frac{L + L_e}{d}\right)_i\right)\Phi_m = 0$$

Equivalent pipe length:	L_e/d
Gate valve, open	8
Ball valve, open	3
90° elbow	30
90° long bend	13
45° elbow	16
45° long bend	10
T connection (through run)	10
T connection (branch run)	60
Hydrodynamic entrance length (laminar flow)	0.05 Re
Hydrodynamic entrance length (turbulent flow)	1.36 Re$^{1/4}$

Heat transfer

Internal versus external heat transfer

$\mathrm{Bi} = \dfrac{h_o d}{\lambda}$ if $\mathrm{Bi} \gg 1$: use internal heat transfer coefficient

if $\mathrm{Bi} \ll 1$: use external heat transfer coefficient

Stationary heat transfer

Heating of turbulent flow in tubes (Re > 4,000, Pr ≥ 0.7):

$$\mathrm{Nu} = 0.027\,\mathrm{Re}^{0.8}\mathrm{Pr}^{0.33} \quad \text{or} \quad \frac{hd}{\lambda} = 0.027 \left(\frac{\rho v d}{\mu}\right)^{0.8} \left(\frac{\mu C_p}{\lambda}\right)^{0.33}$$

Temperature profile for flow through a heated tube

$$\frac{T - T_w}{T_0 - T_w} = \exp\left\{-\frac{2hx}{vR\rho C_p}\right\}$$

Heat transfer from a sphere with forced convection

$$\langle \mathrm{Nu} \rangle = 2.0 + 0.66\,\mathrm{Re}^{0.50}\,\mathrm{Pr}^{0.33} \quad \text{or} \quad \frac{\langle h \rangle d}{\lambda} = 2 + 0.66 \left(\frac{\rho v d}{\mu}\right)^{0.50} \left(\frac{\mu C_p}{\lambda}\right)^{0.33}$$

$$\mathrm{Re} < 10^4,\ \mathrm{Pr} \geq 0.7$$

Heat transfer from a cylinder with forced convection

$$\langle \mathrm{Nu} \rangle = 0.57\,\mathrm{Re}^{0.50}\,\mathrm{Pr}^{0.33} \quad \text{or} \quad \frac{\langle h \rangle d}{\lambda} = 0.57 \left(\frac{\rho v d}{\mu}\right)^{0.50} \left(\frac{\mu C_p}{\lambda}\right)^{0.33}$$

$$10 < \mathrm{Re} < 10^4,\ \mathrm{Pr} \geq 0.7$$

Heat transfer from a plate with forced convection parallel to the plate

$$\langle \mathrm{Nu} \rangle = 0.664\,\mathrm{Re}^{0.5}\,\mathrm{Pr}^{0.33} \quad \text{or} \quad \frac{\langle h \rangle d}{\lambda} = 0.664 \left(\frac{\rho v L}{\mu}\right)^{0.50} \left(\frac{\mu C_p}{\lambda}\right)^{0.33}$$

$$\mathrm{Re} < 3 \times 10^5,\ \mathrm{Pr} \geq 0.7$$

Heat transfer coefficients*: h [W/m² K]

$$h \equiv \frac{\Phi''_h}{\Delta T}$$

Gas, free convection	5–15
Gas, forced convection	10–100
Liquid, free convection	50–1,000
Liquid, forced convection	500–3,000
Condensation of vapor:	1,000–4,000
Boiling of liquid:	1,000–20,000
Water, forced convection:	2,500–15,000
Boiling of water:	3,000–60,000
Condensation of water:	5,000–20,000

*Adapted from Bird R.B., Steward W.E and Lightfoot E.N.: Transport phenomena, John Wiley (1960)

Instationary heat transfer

$Fo = \dfrac{at}{d^2}$ if Fo $\ll$ 0.1: heat penetration

if Fo $>$ 0.1: full body heating

Fo $\ll$ 0.1 $\rightarrow$ Nu $= 0.57\,Fo^{-\frac{1}{2}}$ (all geometries)

Fo $>$ 0.1 $\rightarrow$ Nu $= 4.93$ (plate), Nu $= 5.8$ (cylinder), Nu $= 6.6$ (sphere)

Radiation

$$\Phi''_h = \varepsilon.\sigma\left(T_1^4 - T_2^4\right) \ (T \text{ in Kelvin})$$

$$\sigma = 5.67 \times 10^{-8}\,\mathrm{W/m\cdot K^4} \ \ (\text{Stefan-Boltzmann's constant})$$

Standard normal solar irradiation at ground level:

$$\Phi''_{sol} = 1\,\mathrm{kW/m^2}$$

Emission coefficients (ε):

Polished metals	0.02–0.05
Oxidized metals	0.2–0.6
Brick	0.8–0.9
Concrete	0.6–0.9
Paints, lacquer, oil	0.8–0.9
Paints, aluminum	0.3–0.4
Wood	0.9
Roofing paper	0.9
Human skin	0.97

Heat of combustion of organic compounds * C_c H_h O_o N_n:
$\Delta H_c \approx 418 \cdot (c + 0.3\ h - 0.5\ o)\ kJ/mol$ (if o + n < c)

*Schmidt-Rohr, K.,J. Chem. Educ, 92 (12) p. 2094 (2015)

Mass transfer

$\Phi''_{mol, A} = -\mathcal{D}_A \dfrac{dc_A}{dx}$ (equimolar diffusion)

$\Phi''_{mol, A} = -f_D\, \mathcal{D}_A \dfrac{dc_A}{dx}$ (single-sided diffusion of component A)

$f_D \approx 1 + \dfrac{1}{2} \dfrac{c_{A_1} - c_{A_2}}{\sum_{i=A,B,C,\dots} C_i}$ (Stefan's correction factor for drift flow)

$m = \dfrac{c_{A,i1}}{c_{A,i2}}$ (the Henry – Nernst coefficient)

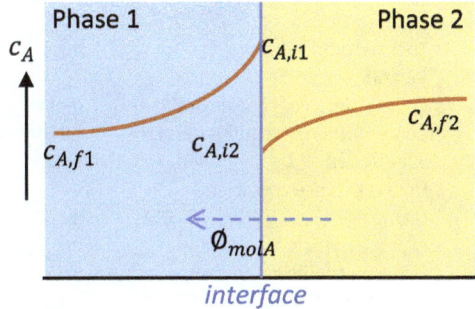

Fig 1.1: concentration jump at an interphase.

$$\Phi''_{mol, A} = \left(\frac{1}{k_2} + \frac{1}{mk_1} \right)^{-1} \left(c_{A,f2} - \frac{c_{A,f1}}{m} \right) \equiv K_2 \left(c_{A,f2} - \frac{c_{A,f1}}{m} \right)$$

or

$$\Phi''_{mol, A} = \left(\frac{m}{k_2} + \frac{1}{k_1} \right)^{-1} \left(m c_{A,f2} - c_{A,f1} \right) \equiv K_1 \left(m c_{A,f2} - c_{A,f1} \right)$$

Diffusion coefficients [m^2/s]*

Gasses in air:	$(10-40) \times 10^{-6}$
Gasses in water:	$(1-5) \times 10^{-9}$
Ions in water:	$(1-2) \times 10^{-9}$
Gas in metal(s):	$(5-500) \times 10^{-14}$
Metal(s) in metal(s):	10^{-34}

*Adapted from Bird R.B., Steward W.E and Lightfoot E.N.: Transport phenomena, John Wiley (1960)

Instationary mass transfer

$$Fo = \frac{\mathcal{D}_A t}{d^2}$$ if Fo $\ll$ 0.1: mass penetration

   if Fo > 0.1: concentration egalization

Fo $\ll$ 0.1 $\rightarrow$ Sh = 0.57 Fo$^{-\frac{1}{2}}$ (all geometries)
Fo > 0.1 $\rightarrow$ Sh = 4.93 (plate), Sh = 5.8 (cylinder), Sh = 6.6 (sphere)

List of symbols and abbreviations

a	Thermal diffusivity	m^2/s	m	Henry-Nernst coefficient	–
Bi	Biot number	–	Nu	Nusselt number	–
C_A	Concentration of component A	mol/m^3	<Nu>	Average Nusselt number	–
$C_{A,f}$	Concentration of A at the interface	mol/m^3	p	Pressure	Pa
C_p	Specific heat	kJ/kg·K	Pr	Prandtl number	–
C_w	Flow resistance coefficient	–	R	Radius	m
d	Diameter	m	Re	Reynolds number	–
d_{30}	Average particle diameter	m	T	Temperature	K or °C
$\mathcal{D}_A$	Diffusion coefficient of component A	m^2/s	T_0	Initial temperature	K or °C
f	Friction coefficient	–	T_w	Wall temperature	K or °C
F	Force	N	v	Velocity	m/s
f_D	Correction factor of Stefan	–	x	Coordinate, distance	m
Fo	Fourier number		ε	Emission coefficient	–
g	Gravitational coefficient	m/s^2	λ	Thermal conductivity	W/m·K
h	Height	m	μ	Viscosity	Pa·s
h_0	External heat transfer coefficient	W/m·K	ρ	Density	kg/m^3
H_c	Heat of combustion	kJ/mol	σ	Stefan-Boltzmann's constant	W/m·K^4
k	Mass transfer coefficient	m/s	Φ''_h	Heat flux	W/m^2
K	Overall mass transfer coefficient	m/s	$\Phi''_{mol,A}$	Molar flux of component A	mol/m^{2}s
k_p	Permeability	m^2	Φ''_{sol}	Insolation of solar energy	W/m^2
L	Length	m	Φ_A	Energy added	W
L_e	Equivalent length	m	Φ_m	Mass flow	kg/s
Sh	Sherwood number	–	<Sh>	Average Sherwood number	–

Chapter 2
Miscellaneous

Greek alphabet

A	α	Alpha	N	ν	Nu
B	β	Beta	Ξ	ξ	Xi
Γ	γ	Gamma	O	o	Omicron
Δ	δ	Delta	Π	π	Pi
E	ϵ	Epsilon	P	ρ	Rho
Z	ζ	Zeta	Σ	σ	Sigma
H	η	Eta	T	τ	Tau
Θ	ϑ	Theta	Y	υ	Upsilon
I	ι	Iota	Φ	φ	Phi
K	κ	Kappa	X	χ	Chi
Λ	λ	Lambda	Ψ	ψ	Psi
M	μ	Mu	Ω	ω	Omega

Système Internationale (International System of Units)

SI base units

Quantity	Dimension symbol	SI unit	Symbol SI unit
Length	L	Meter	m
Mass	M	Kilogram	kg
Time	T	Second	s
Temperature	θ	Kelvin	K
Amount of substance	N	Mole	mol
Luminous intensity	J	Candela	cd
Electric current	I	Ampere	A

SI named units*

Name	Symbol	Quantity	In derived SI units	In SI base units
Radian	rad	Plane angle	m/m	–
Steradian	sr	Solid angle	m^2/m^2	–

https://doi.org/10.1515/9783111385341-002

(continued)

Name	Symbol	Quantity	In derived SI units	In SI base units
Hertz	Hz	Frequency	1/s	s^{-1}
Newton	N	Force, weight	$kg \cdot m/s^2$	$kg \cdot m \cdot s^{-2}$
Pascal	Pa	Pressure, stress	N/m^2	$kg \cdot m^{-1} \cdot s^{-2}$
Joule	J	Energy, work, heat	$N \cdot m = Pa \cdot m^3$	$kg \cdot m^2 \cdot s^{-2}$
Watt	W	Power, radiant flux	J/s	$kg \cdot m^2 \cdot s^{-3}$
Coulomb	C	Electric charge	s·A	s·A
Volt	V	Electrical potential, voltage, emf	W/A = J/C	$kg \cdot m^2 \cdot s^{-3} \cdot A^{-1}$
Farad	F	Capacitance	C/V	$kg^{-1} \cdot m^{-2} \cdot s^4 \cdot A^2$
Ohm	Ω	Resistance, impedance, reactance	V/A	$kg \cdot m^2 \cdot s^{-3} \cdot A^{-2}$
Siemens	S	Electrical conductance	1/Ω	$kg^{-1} \cdot m^{-2} \cdot s^3 \cdot A^2$
Weber	Wb	Magnetic flux	V·s	$kg \cdot m^2 \cdot s^{-2} \cdot A^{-1}$
Tesla	T	Magnetic flux density	Wb/m^2	$kg \cdot s^{-2} \cdot A^{-1}$
Henry	H	Inductance	Wb/A	$kg \cdot m^2 \cdot s^{-2} \cdot A^{-2}$
Degree Celsius	°C	Temperature relative to 273.15 K	K	K
Lumen	lm	Luminous flux	cd·sr	cd
Lux	lx	Illuminance	lm/m^2	$m^{-2} \cdot cd$
Becquerel	Bq	Radiation activity (decays per unit time)	1/s	s^{-1}
Gray	Gy	Absorbed dose (of ionizing radiation)	J/kg	$m^2 \cdot s^{-2}$
Sievert	Sv	Equivalent dose (of ionizing radiation)	J/kg	$m^2 \cdot s^{-2}$
Katal	kat	Catalytic activity	mol/s	$mol \cdot s^{-1}$

SI accepted units*

Quantity	Name	Symbol	Value in SI units
Time	Minute	min	1 min = 60 s
	Hour	h	1 h = 60 min = 3,600 s
	Day	d	1 d = 24 h = 86,400 s
Length	Astronomical unit	au	1 au = 149,597,870,700 m
Plane and phase angle	Degree	°	$1° = (\pi/180)$ rad
	Minute	′	$1' = (1/60)° = (\pi/10, 800)$ rad
	Second	″	$1'' = (1/60)' = (\pi/648, 000)$ rad
Area	Hectare	ha	$1\,ha = 1\,hm^2 = 10^4\ m^2$
Volume	Liter	l, L	$1\,l = 1\,L = 1\,dm^3 = 10^3\,cm^3 = 10^{-3}\ m^3$
Mass	Ton (metric)	t	1 t = 1,000 kg
	Dalton	Da	$1\,Da = 1.660539040(20) \times 10^{-27}$ kg
Energy	Electron volt	eV	$1\,eV = 1.602176634 \times 10^{-19}$ J

*Adapted from The BIPM and the Metre Convention, S.I. Brochure, Bureau International des Poids et Mesures (2019).

SI prefixes*

Name	Symbol	Base 10	Decimal	Short scale	Long scale
Yotta	Y	10^{24}	1,000,000,000,000,000,000,000,000	Septillion	Quadrillion
Zetta	Z	10^{21}	1,000,000,000,000,000,000,000	Sextillion	Trilliard
Exa	E	10^{18}	1,000,000,000,000,000,000	Quintillion	Trillion
Peta	P	10^{15}	1,000,000,000,000,000	Quadrillion	Billiard
Tera	T	10^{12}	1,000,000,000,000	Trillion	Billion
Giga	G	10^{9}	1,000,000,000	Billion	Milliard
Mega	M	10^{6}	1,000,000	Million	
Kilo	k	10^{3}	1,000	Thousand	
Hecto	h	10^{2}	100	Hundred	
Deca	da	10^{1}	10	Ten	
		10^{0}	1	One	
Deci	d	10^{-1}	0.1	Tenth	
Centi	c	10^{-2}	0.01	Hundredth	
Milli	m	10^{-3}	0.001	Thousandth	
Micro	μ	10^{-6}	0.000001	Millionth	
Nano	n	10^{-9}	0.000000001	Billionth	Milliardth
Pico	p	10^{-12}	0.000000000001	Trillionth	Billionth
Femto	f	10^{-15}	0.000000000000001	Quadrillionth	Billiardth
Atto	a	10^{-18}	0.000000000000000001	Quintillionth	Trillionth
Zepto	z	10^{-21}	0.000000000000000000001	Sextillionth	Trilliardth
Yocto	y	10^{-24}	0.000000000000000000000001	Septillionth	Quadrillionth

Periodic table

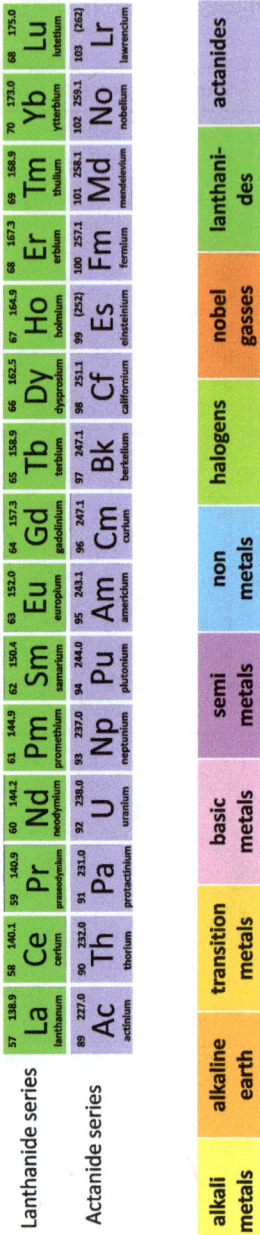

Fig. 2.1: Periodic system of elements.

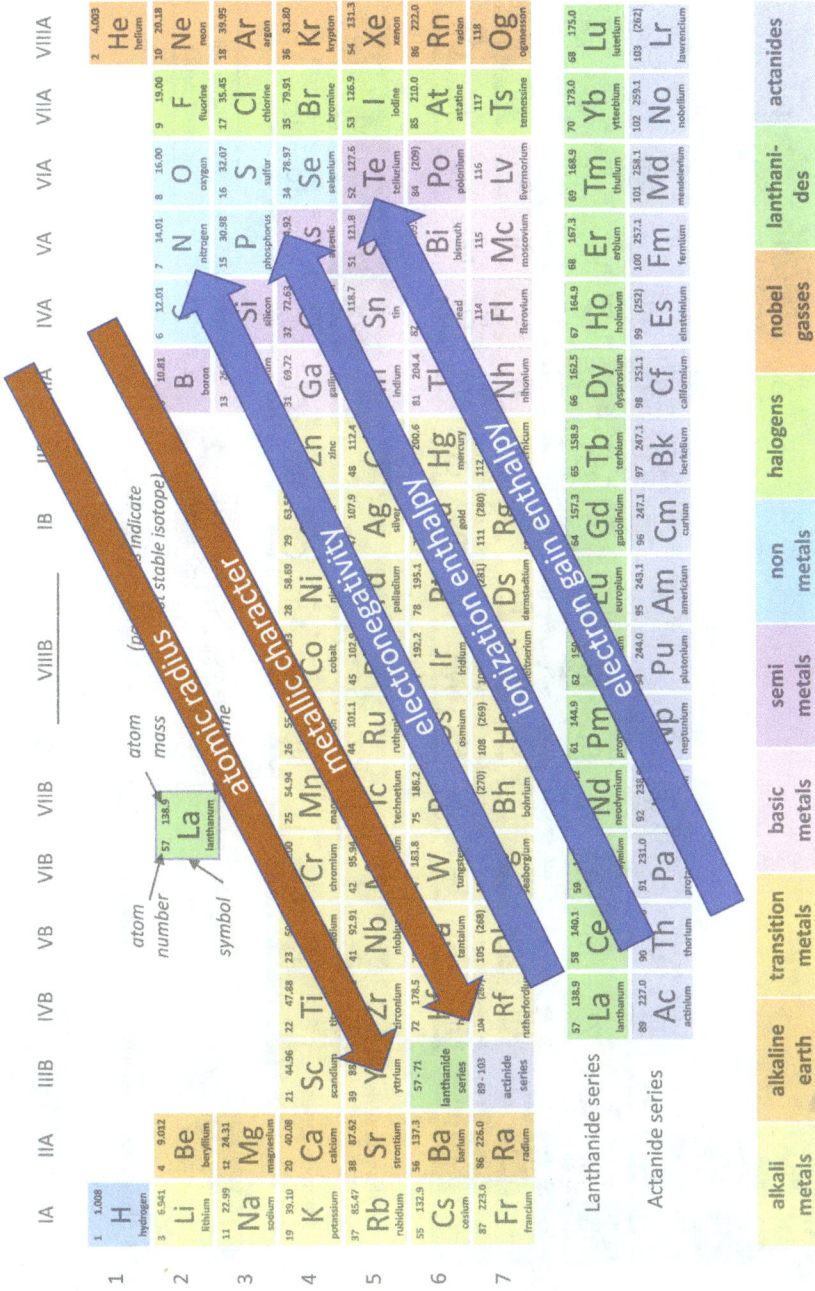

Fig. 2.2: Trends in the periodic system.

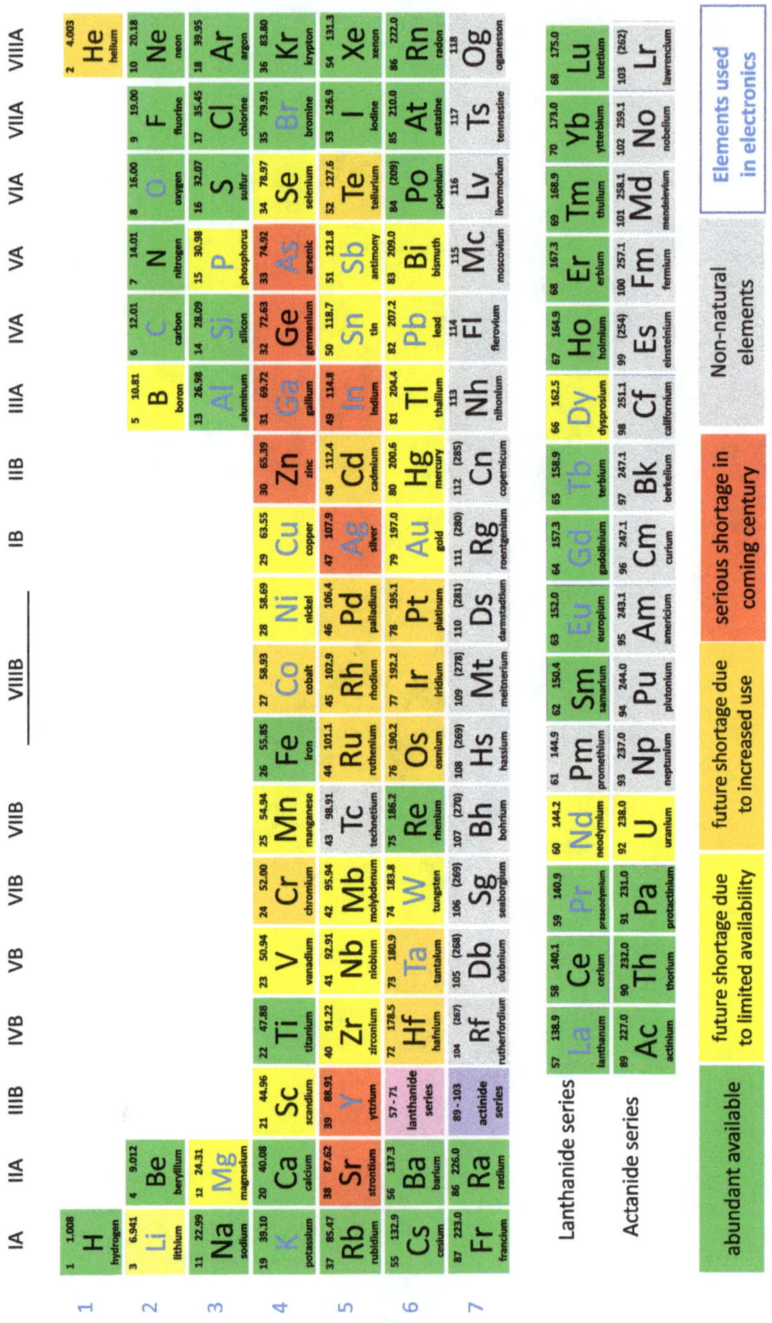

Fig. 2.3: Abundance of elements. Adapted from: Chemistry Innovation Knowledge Transfer Network.

Mathematical constants

π	=	3.141592653590
e	=	2.718281828459
π^e	=	22.459157718361
e^π	=	23.140692632779
$^{10}\log e$	=	0.434294481903
$^e\log 10$	=	2.302585092994
1 radian	=	57° 17′ 44.8″
1°	=	0.0174532925 radians
1′	=	0.0002908882 radians
1″	=	0.0000048481 radians

Physical constants

Gas constant	R	8.314413	J/mol K
Avogadro's number	N	6.022141×10^{23}	1/mol
Gravitational acceleration	g		
Standard		9.80665	m/s^2
Pole		9.81422	m/s^2
Equator		9.78039	m/s^2
Volume of an ideal gas	V_m	2.24138×10^{-2}	m^3/mol
(at standard pressure and temperature)			
Stefan-Boltzmann's constant	σ	5.67037×10^{-8}	W/m^2 K^4
Wien's displacement constant	b	2.898×10^{-3}	m K
Boltzmann's constant	k	1.380649×10^{-23}	J/K
Standard temperature	T_s	273.15	K
Standard pressure .	P_s	1.01325×10^5	Pa
Speed of light	c	2.998×10^8	m/s

Values of the gas constant $P \times V = n\,R\,T$ $(n \text{ in mol})$

Temperature	K (kelvin)			R (Rankine)		
Volume	Liter	m^3	ft^3	Liter	m^3	ft^3
Pressure						
Pa	8.314	8.314	294	4620	4.62	163
atm.	0.08205	8.205×10^{-5}	0.00290	0.0456	4.56×10^{-5}	0.00161
cm H_2O	84.79	0.08479	2.99	47.1	0.0471	1.66
mm Hg	62.4	0.0624	2.20	34.6	0.0346	1.22

Example: What is the volume in liters of 2 mol of gas at 30 mm Hg and 20 °C?

$$V = \frac{n\,R\,T}{P} = \frac{2 \times 62.4 \times (20 + 273.15)}{30} = 1{,}220 \text{ liters}$$

Conversion factors

Symbol	Quantity	Expressed in	Multiply by ⟶ ⟵ Divided by	In SI units (derived units)
L	Length	Inch (in)	0.0254	m
		Foot (ft)	0.3048	
		Yard (yd)	0.9144	
		Mile	1609	
		Ångstrom (Å)	10^{-10}	
A	Surface	in^2	6.45×10^{-4}	m^2
		ft^2	0.0929	
		yd^2	0.836	
		Acre	4047	
		$Mile^2$	2.59×10^6	
V	Volume	in^3	1.64×10^{-5}	m^3
		ft^3	0.0283	
		yd^3	0.765	
		UK gallon	4.55×10^{-3}	
		US gallon	3.785×10^{-3}	
t	Time	Minute	60	s
		Hour	3,600	
		Day	8.64×10^4	
		Year	3.16×10^7	
m	Mass	Grain	6.48×10^{-5}	kg
		Ounce (oz)	2.84×10^{-2}	
		Ounce (troy)	3.11×10^{-2}	
		Pound (lb)	0.454	
		Hundredweight (cwt)	50.8	
		Ton (long)	1016	
		Ton (short)	907.18	
		Ton (metric)	1000	
F	Force	Poundal (pdl)	0.138	N
		Pound force (lbf)	4.45	$(kg\ m/s^2)$
		Dyn	10^{-5}	
		kg force	9.81	
φ_v	Flow rate (volume)	ft^3/min	4.72×10^{-4}	m^3/s
		UK gal/min	7.58×10^{-5}	
		US gal/min	6.31×10^{-5}	
φ_m	Flow rate (mass)	lb/min	7.56×10^{-3}	kg/s
		ton/h	0.282	
ρ	Density	lb/in^3	2.77×10^4	kg/m^3
		lb/ft^3	16.0	

(continued)

p	Pressure	lbf/in^2 (= psi)	6.89×10^3	$Pa = N/m^2$
		lbf/ft^2	47.9	(kg/ms^2)
		dyn/cm^2	0.1	
		kgf/cm^2 (= at)	9.81×10^4	
		atm (standard)	1.013×10^5	
		bar	10^5	
		inch water	2.49×10^2	
		mm water	9.80	
		ft water	2.99×10^3	
		inch Hg	3.39×10^3	
		mm Hg (torr)	1.33×10^2	
η	Dynamic viscosity	lb/ft h	4.13×10^{-4}	$Pa\ s = Ns/m^2$
		lb/ft s	1.49	(kg/ms)
		poise (P = g/cm s)	0.1	
		centipoise (cP)	10^{-3}	
ν	Kinematic viscosity	ft^2/h	2.58×10^{-5}	m^2/s
		Stokes (S = cm^2/s)	10^{-4}	
		Centistokes (cS)	10^{-6}	
σ	Surface tension	dyn/cm (= erg/cm^2)	10^{-3}	$N/m\ (kg/s^2)$
ΔT	Temperature difference	Degree F (or R)	5/9	K (°C)
Q	Energy, work, amount of heat	ft lbl	0.0421	$J = W\ s = Nm$
		ft lbf	1.36	(kgm^2/s^2)
		BTU	1.06×10^3	
		CHU	1899	
		hph	2.68×10^6	
		erg	10^{-7}	
		kgf m	9.81	
		kcal	4.19×10^3	
		kWh	3.60×10^6	
P	Power, energy flux	BTU/h	0.293	$W = J/s = Nm/s$
		CHU/h	0.528	(kgm^2/s^3)
		ft lbf/s	1.36	
		hp (British)	746	
		hp (metric)	736	
		erg/s	10^{-7}	
		kcal/h	1.163	
		cal/s	4.19	
φ_w''	Heat flux	$BTU/ft^2\ h$	3.15	W/m^2
		$cal/cm^2\ s$	4.19×10^4	(kg/s^3)
		$kcal/m^2\ h$	1.163	

(continued)

c	Specific heat	BTU/lb °F	4.19×10^3	J/kg K
		kcal/kg °C	4.19×10^3	$(m^2/s^2\,K)$
	Latent heat	BTU/lb	2.33×10^3	J/kg
		kcal/kg	4.19×10^3	(m^2/s^2)
λ	Heat conductivity	BTU/ft h °F	1.73	W/m K
		cal/cm s °C	4.19×10^2	$(kg\,m/s^3)$
		kcal/m h °C	1.163	
h	Heat transfer coefficient	BTU/ft^2 h °F	5.68	W/m^2 K
		cal/cm^{2}s °C	4.19×10^4	$(kg/s^3\,K)$
		kcal/m^2 h °C	1.163	

Temperature conversion

From \ To	K	°C	°F	R
Kelvin	1	$K - 273.15$	$\frac{9}{5}K - 459.69$	$\frac{9}{5}K$
°Celsius	$°C + 273.15$	1	$\frac{9}{5}°C + 32$	$\frac{9}{5}°C + 491.67$
°Fahrenheit	$\frac{5}{9}°F + 255.37$	$\frac{5}{9}°F - 17.78$	1	$°F + 459.7$
Rankine	$\frac{5}{9}R$	$\frac{5}{9}R - 273.15$	$R - 459.7$	1

Temperature scales

Temperature in	Temperature of melting ice	Temperature of boiling water
Celsius	0 °C	100 °C
Rankine	491.67 R	671.67 R
Fahrenheit	32 °F	212 °F
Kelvin	273.15 K	373.15 K

Approximations for physical properties of water and air

Notation

C	Specific heat	kJ/kg K
p	Pressure	Pa
T	Temperature (0 °C < T < 100 °C)	°C
η	Viscosity	Pa s
λ	Heat conductivity	W/m K
ρ	Density	kg/m^3
H_a	Absolute humidity	–
H_r	Relative humidity	–
X	Mole fraction	–
$\mathbb{D}$	Binary diffusion coefficient	m^2/s
M	Molar mass	kg/mol

Indices

d	Dry air
h	Per unit mass of dry air
n	Net dry air plus contribution of the water vapor
s	Air saturated with water
v	Water vapor

Water

$$\eta = 10^{-3} \times \exp\left(0.580 - 2.520\ \Theta + 0.909\ \Theta^2 - 0.264\ \Theta^3\right) \text{ Pa s}$$
$$\text{with } \Theta = 3.6610\ T/(273.15 + T)$$
$$\lambda = 0.5607 + 0.200\ (T/100) - 0.795\ (T/100)^2 \text{ W/m K}$$

Water vapor

$$M_v = 18.02 \times 10^{-3} \qquad \text{kg/mol}$$
$$C_v = 1.87 \qquad \text{kJ/kg K}$$
$$\rho_v = \frac{2.165 \times 10^{-3} \, P}{273.15 + T} \qquad \text{kg/m}^3$$
$$\eta_v = \left(8.19 + 0.0407 \, T + 4.73 \times 10^{-7} \, P\right) \times 10^{-6} \qquad \text{Pa s}$$
$$\lambda_v = 0.0158 + 84 \times 10^{-6} \, T \qquad \text{W/m K}$$

Humid air

$$H_a = 0.622 \, p_v / (p_n - p_v) \qquad -$$
$$H_r = p_v / p_{vs} \qquad -$$
$$x_v = H_a / (0.622 + H_a) \qquad -$$
$$\rho_n = (3.484 - 1.317 \, x_v) p_n / (273.15 + T) \qquad \text{kg/m}^3$$
$$\eta_n = (\eta_d + y_v(0.7887 \, \eta_v - \eta_d)) / (1 - 0.2113 \, x_v) \times 10^{-6} \qquad \text{Pa s}$$
$$\lambda_n = (\lambda_d + x_v(0.8536 \, \lambda_v - \lambda_d)) / (1 - 0.1464 \, x_v) \times 10^{-3} \qquad \text{W/m K}$$
$$C_h = C_d + H_a C_v \qquad \text{kJ/kg K}$$
$$\mathbb{D} = 2.22 \, (1 + T/273.15)^{1.75} / p_n \qquad \text{m}^2/\text{s}$$

Dry air

$$M_d = 28.97 \times 10^{-3} \qquad \text{kg/mol}$$
$$C_d = 1.00 \qquad \text{kJ/kg K}$$
$$\rho_d = 3.484 \times 10^{-3} p_d / (273.15 + T) \qquad \text{kg/m}^3$$
$$\eta_d = \left(17.11 + 0.0536 \, T + 9.058 \times 10^{-7} \, p\right) 10^{-6} \qquad \text{Pa s}$$
$$\lambda = 0.02402 + 74 \times 10^{-6} \, T \qquad \text{W/m K}$$

pH range of acid-base indicators

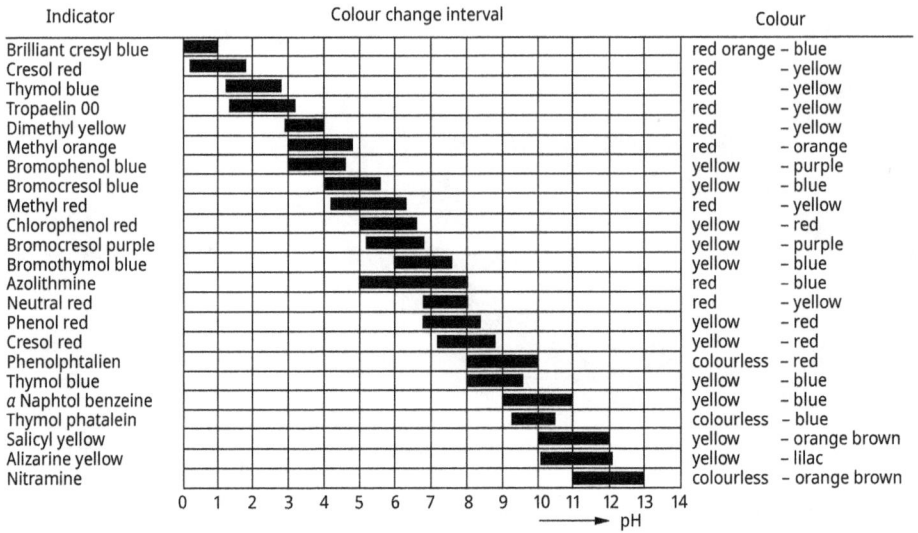

Indicator	Colour change interval	Colour
Brilliant cresyl blue		red orange – blue
Cresol red		red – yellow
Thymol blue		red – yellow
Tropaelin 00		red – yellow
Dimethyl yellow		red – yellow
Methyl orange		red – orange
Bromophenol blue		yellow – purple
Bromocresol blue		yellow – blue
Methyl red		red – yellow
Chlorophenol red		yellow – red
Bromocresol purple		yellow – purple
Bromothymol blue		yellow – blue
Azolithmine		red – blue
Neutral red		red – yellow
Phenol red		yellow – red
Cresol red		yellow – red
Phenolphtalien		colourless – red
Thymol blue		yellow – blue
α Naphtol benzeine		yellow – blue
Thymol phatalein		colourless – blue
Salicyl yellow		yellow – orange brown
Alizarine yellow		yellow – lilac
Nitramine		colourless – orange brown

pH scale: 0 1 2 3 4 5 6 7 8 9 10 11 12 13 14 → pH

Fig. 2.4: colour change of indicators.

Chapter 3
Mathematics

Quadratic and cubic relations

$$(x + a)(x + b) = x^2 + (a + b)x + ab$$

$$(a + b)^2 = a^2 + 2ab + b^2$$

$$(a - b)^2 = a^2 - 2ab + b^2$$

$$(a + b)(a - b) = a^2 - b^2$$

$$(a + b)^2 = (a - b)^2 + 4ab$$

$$(a - b)^2 = (a + b)^2 - 4ab$$

$$(a + b)^2 - (a - b)^2 = 4ab$$

$$(a + b)^2 + (a - b)^2 = 2(a^2 + b^2)$$

$$(a + b + c)^2 = a^2 + b^2 + c^2 + 2ab + 2bc + 2ac$$

$$(a + b)^3 = a^3 + 3a^2b + 3ab^2 + b^3$$

$$(a + b)^3 = a^3 + b^3 + 3ab(a + b)$$

$$(a - b)^3 = a^3 - 3a^2b + 3ab^2 - b^3$$

$$(a - b)^3 = a^3 - b^3 - 3ab(a - b)$$

$$a^3 + b^3 = (a + b)(a^2 - ab + b^2)$$

$$a^3 - b^3 = (a - b)(a^2 + ab + b^2)$$

$$(a + b + c)(a^2 + b^2 + c^2 - ab - bc - ac) = a^2 + b^2 + c^2 - 3abc$$

https://doi.org/10.1515/9783111385341-003

Volume and surface of some bodies

V = volume	A = surface
S = surface	h = height
S_s = surface sides	

Cube	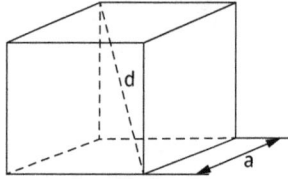	$S = 6a^2$ $V = a^3$ $d = a\sqrt{3}$
Block	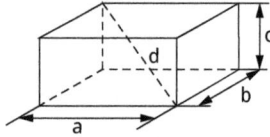	$S = 2\,(ab + ac + bc)$ $V = a \cdot b \cdot c$ $d = \sqrt{a^2 + b^2 + c^2}$
Pyramid	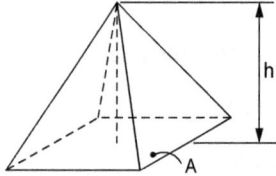	$V = \dfrac{1}{3}Ah$
Tetrahedron	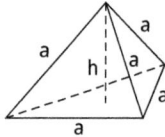	$V = \dfrac{1}{12}a^3\sqrt{2}$ $S = a^2\sqrt{3}$ $h = \dfrac{1}{3}a\sqrt{6}$
Octahedron	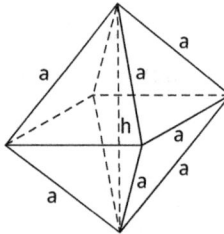	$V = \dfrac{1}{3}a^3\sqrt{2}$ $S = 2a^2\sqrt{3}$ $h = a\sqrt{2}$
Truncated pyramid	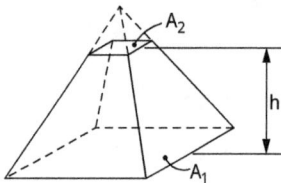	$V = \dfrac{1}{3}h(A_1 + A_2 + \sqrt{A_1 A_2})$

(continued)

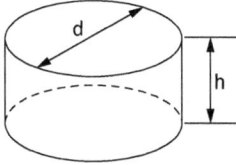

Cylinder		$V = \frac{1}{4}\pi h d^2$
		$S = \pi d\left(\frac{1}{2}d + h\right)$
		$S_s = \pi dh$

Cylinder

$$V = \frac{1}{4}\pi h d^2$$
$$S = \pi d\left(\frac{1}{2}d + h\right)$$
$$S_s = \pi dh$$

Tube

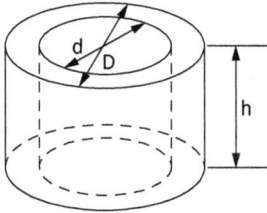

$$V = \frac{1}{4}\pi h\left(D^2 - d^2\right)$$

Truncated cylinder

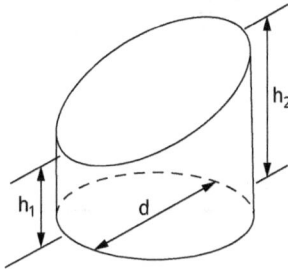

$$V = \frac{1}{8}\pi d^2\left(h_1 + h_2\right)$$
$$S = \frac{1}{2}\pi d\left[h_1 + h_2 + \frac{1}{2}d + \right.$$
$$\left. + \sqrt{\frac{1}{4}d^2 + \left(\frac{h_1 - h_2}{2}\right)^2}\right]$$
$$S_s = \frac{1}{2}\pi d(h_1 + h_2)$$

Cylinder segment

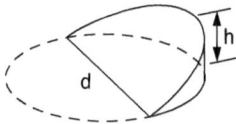

$$V = \frac{1}{6}hd^2$$

Cone

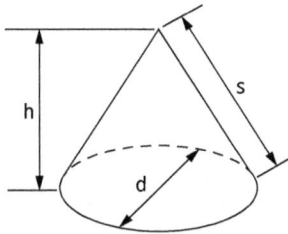

$$V = \frac{1}{12}\pi d^2 h$$
$$S = \frac{1}{2}\pi d\left(\frac{1}{2}d + s\right)$$
$$S_s = \frac{1}{2}\pi ds$$

Truncated cone

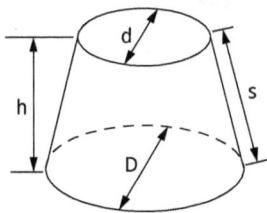

$$V = \frac{1}{12}\pi h\left(D^2 + Dd + d^2\right)$$
$$S = \frac{1}{2}\pi D\left(\frac{1}{2}D + s\right) + \frac{1}{2}\pi d\left(\frac{1}{2}d + s\right)$$

(continued)

Sphere	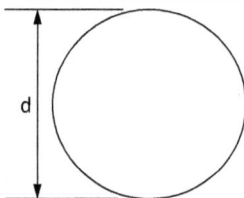	$V = \frac{1}{6}\pi d^3$ $S = \pi d^2$
Sphere segment	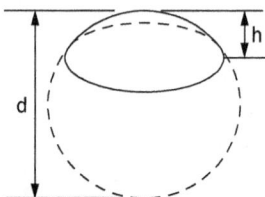	$V = \pi h^2 \left(\frac{1}{2}d - \frac{1}{3}h \right)$ $S = \pi \left(2dh - h^2 \right)$
Sphere sector	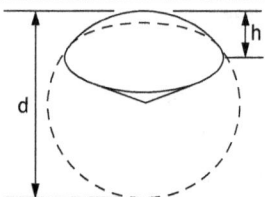	$V = \frac{1}{6}\pi d^2 h$ $S = \frac{1}{2}\pi d \left(2h + \sqrt{hd - h^2} \right)$

Integrals

To all solutions, an integration constant has to be added.
u and v are functions of x.

Basic integrals

$$\int a \; du = au$$

$$\int \frac{du}{u} = \ln u$$

$$\int u^n \; du = \frac{u^{n+1}}{n+1}$$

$$\int e^u \; du = e^u$$

$$\int \sin u \; du = -\cos u$$

$$\int \tan u \; du = -\ln \cos u$$

$$\int \csc^2 u \; du = -\cot u$$

$$\int \cot u \; \csc u \; du = -\csc u$$

$$\int \csc u \; du = \ln(\csc u - \cot u)$$

$$\int a^u \; du = \frac{a^u}{\ln a}$$

$$\int \frac{du}{a^2 + u^2} = \frac{1}{a} \arctan \frac{u}{a} \text{ of } -\frac{1}{a} \operatorname{arccot} \frac{u}{a}$$

$$\int \frac{du}{\sqrt{a^2 - u^2}} = \arcsin \frac{u}{a} \text{ of } -\arccos \frac{u}{a}$$

$$\int \cos u \; du = \sin u$$

$$\int \cot u \; du = \ln \sin u$$

$$\int \sec^2 u\ du = \tan\ u$$

$$\int \tan\ u\ \sec\ u\ du = \sec\ u$$

$$\int \sec\ u\ du = \ln(\sec\ u + \tan\ u)$$

$$\int (u+v)\ dx = \int u\ dx + \int v\ dx$$

$$\int u\ dv = uv - \int v\ du$$

Functions of *a + bx*

$$\int (a+bx)^n dx = \frac{1}{b(n+1)}(a+bx)^{n+1} \qquad (n \neq -1)$$

$$\int x(a+bx)^n dx = \frac{1}{b^2(n+2)}(a+bx)^{n+2} - \frac{a}{b^2(n+1)}(a+bx)^{n+1}$$

or

$$\frac{x}{b(n+1)}(a+bx)^{n+1} - \frac{1}{b^2(n+1)(n+2)}(a+bx)^{n+2} \qquad (n \neq -1, -2)$$

$$\int x^2(a+bx)^n dx = \frac{(a+bx)^{n+1}}{b^3}\left[\frac{(a+bx)^2}{n+3} - \frac{2a(a+bx)}{(n+2)} + \frac{a^2}{(n+1)}\right] \qquad (n \neq -1, -2, -3)$$

$$\int \frac{dx}{a+bx} = \frac{1}{b}\ln(a+bx)$$

$$\int \frac{dx}{(a+bx)^2} = -\frac{1}{b}\frac{1}{a+bx}$$

$$\int \frac{x\ dx}{a+bx} = \frac{x}{b} - \frac{a}{b^2}\ln(a+bx)$$

$$\int \frac{x^2\ dx}{a+bx} = \frac{1}{b^3}\left[\frac{1}{2}b^2x^2 - abx + a^2\ln(a+bx)\right]$$

$$\int \frac{x\ dx}{(a+bx)^2} = \frac{a}{b^2(a+bx)} + \frac{1}{b^2}\ln(a+bx)$$

$$\int \frac{x^2\,dx}{(a+bx)^2} = \frac{1}{b^3}\left[a+bx - \frac{a^2}{a+bx} - 2a\ln(a+bx)\right]$$

$$\int \frac{dx}{x(a+bx)} = \frac{1}{a}\ln\frac{x}{a+bx}$$

$$\int \frac{dx}{x^2(a+bx)} = -\frac{1}{ax} + \frac{b}{a^2}\ln\frac{a+bx}{x}$$

$$\int \frac{dx}{x(a+bx)^2} = -\frac{1}{a(a+bx)} - \frac{1}{a^2}\ln\frac{a+bx}{x}$$

$$\int \frac{dx}{x^2(a+bx)^2} = -\frac{a+2bx}{a^2x(a+bx)} + \frac{2b}{a^3}\ln\frac{a+bx}{x}$$

Functions of $a + bx^2$

$$\int \frac{dx}{a+bx^2} = \frac{1}{\sqrt{ab}}\arctan\left(x\sqrt{\frac{b}{a}}\right)\quad (a,b>0)$$

$$= \frac{1}{2\sqrt{-ab}}\ln\frac{\sqrt{a}+x\sqrt{-b}}{\sqrt{a}-x\sqrt{-b}}\quad (a>0,\ b<0)$$

$$\int \frac{dx}{(a+bx^2)^2} = \frac{x}{2a(a+bx^2)} + \frac{1}{2a}\int\frac{dx}{a+bx^2}$$

$$\int \frac{xdx}{a+bx^2} = \frac{1}{2b}\ln\left(x^2+\frac{a}{b}\right)$$

$$\int \frac{dx}{x(a+bx^2)} = \frac{1}{2a}\ln\frac{x^2}{a+bx^2}$$

$$\int \frac{dx}{x^2(a+bx^2)} = -\frac{1}{ax} - \frac{b}{a}\int\frac{dx}{a+bx^2}$$

Functions of $a + bx + cx^2$

$$\int \frac{dx}{a+bx+cx^2} = \frac{1}{\sqrt{b^2-4ac}}\ln\frac{2cx+b-\sqrt{b^2-4ac}}{2cx+b+\sqrt{b^2-4ac}}\quad (b^2>4ac)$$

$$\int \frac{dx}{a+bx+cx^2} = \frac{2}{\sqrt{4ac-b^2}}\arctan\frac{2cx+b}{\sqrt{4ac-b^2}}\quad (b^2<4ac)$$

$$\int \frac{dx}{a+bx+cx^2} = -\frac{2}{2cx+b} \quad (b^2 = 4ac)$$

$$\int \frac{xdx}{a+bx+cx^2} = \frac{1}{2c}\ln(a+bx+cx^2) - \frac{b}{2c}\int \frac{dx}{a+bx+cx^2}$$

$$\int \frac{x^2 dx}{a+bx+cx^2} = \frac{x}{c} - \frac{b}{2c^2}\ln(a+bx+cx^2) + \frac{b^2-2ac}{2c^2}\int \frac{dx}{a+bx+cx^2}$$

Functions of $\sqrt{a+bx}$

$$\int \sqrt{a+bx}\ dx = \frac{2}{3b}(a+bx)^{\frac{3}{2}}$$

$$\int x\sqrt{a+bx}\ dx = \frac{2(3bx-2a)(a+bx)^{\frac{3}{2}}}{15b^2}$$

$$\int x^2\sqrt{a+bx}\ dx = \frac{2(8a^2-12abx+15b^2x^2)(a+bx)^{\frac{3}{2}}}{105b^3}$$

$$\int \frac{\sqrt{a+bx}}{x}\ dx = 2\sqrt{a+bx} + a\int \frac{dx}{x\sqrt{a+bx}}$$

$$\int \frac{dx}{\sqrt{a+bx}} = \frac{2\sqrt{a+bx}}{b}$$

$$\int \frac{x\ dx}{\sqrt{a+bx}} = \frac{2(bx-2a)}{3b^2}\sqrt{a+bx}$$

$$\int \frac{x^2\ dx}{\sqrt{a+bx}} = \frac{2(8a^2-4abx+3b^2x^2)}{15b^3}\sqrt{a+bx}$$

$$\int \frac{dx}{x^2\sqrt{a+bx}} = -\frac{\sqrt{a+bx}}{ax} - \frac{b}{2a}\int \frac{dx}{x\sqrt{a+bx}}$$

$$\int \frac{dx}{x\sqrt{a+bx}} = \frac{1}{\sqrt{a}}\ln\frac{\sqrt{a+bx}-\sqrt{a}}{\sqrt{a+bx}+\sqrt{a}} \quad (a>0)$$

$$\int \frac{dx}{x\sqrt{a+bx}} = \frac{2}{\sqrt{-a}}\arctan\frac{\sqrt{a+bx}}{-a} \quad (a<0)$$

Functions of $\sqrt{a^2+x^2}$

$$\int \sqrt{a^2+x^2}\; dx = \frac{1}{2}\left[x\sqrt{a^2+x^2} + a^2\ln\left(x + \sqrt{a^2+x^2}\right)\right]$$

$$\int \frac{dx}{\sqrt{a^2+x^2}} = \ln\left(x + \sqrt{a^2+x^2}\right)$$

$$\int \frac{dx}{x\sqrt{a^2+x^2}} = -\frac{1}{a}\ln\left(\frac{a + \sqrt{a^2+x^2}}{x}\right)$$

$$\int \frac{\sqrt{a^2+x^2}}{x}\; dx = \sqrt{a^2+x^2} - a\ln\left(\frac{a + \sqrt{a^2+x^2}}{x}\right)$$

$$\int \frac{x\; dx}{\sqrt{a^2+x^2}} = \sqrt{a^2+x^2}$$

$$\int x\sqrt{a^2+x^2}\; dx = \frac{1}{3}\left(a^2+x^2\right)^{\frac{3}{2}}$$

Functions of $\sqrt{a^2-x^2}$

$$\int \sqrt{a^2-x^2}\; dx = \frac{1}{2}\left(x\sqrt{a^2-x^2} + a^2\arcsin\frac{x}{a}\right)$$

$$\int \frac{dx}{x\sqrt{a^2-x^2}} = -\frac{1}{a}\ln\left(\frac{a + \sqrt{a^2-x^2}}{x}\right)$$

$$\int \frac{\sqrt{a^2-x^2}}{x}\; dx = \sqrt{a^2-x^2} - a\ln\left(\frac{a + \sqrt{a^2-x^2}}{x}\right)$$

$$\int \frac{x\; dx}{\sqrt{a^2-x^2}} = -\sqrt{a^2-x^2}$$

$$\int x\sqrt{a^2-x^2}\; dx = -\frac{1}{3}\left(a^2-x^2\right)^{\frac{3}{2}}$$

$$\int \left(a^2-x^2\right)^{\frac{3}{2}} dx = \frac{1}{4}\left[x(a^2-x^2)^{\frac{3}{2}} + \frac{3a^2x}{2}\sqrt{a^2-x^2} + \frac{3a^4}{2}\arcsin\frac{x}{a}\right]$$

$$\int x^2\sqrt{a^2-x^2}\; dx = -\frac{x}{4}(a^2-x^2)^{\frac{3}{2}} + \frac{a^2}{8}\left(x\sqrt{a^2-x^2} + a^2\arcsin\frac{x}{a}\right)$$

$$\int \frac{x^2 dx}{\sqrt{a^2-x^2}} = -\frac{x}{2}\sqrt{a^2-x^2} + \frac{a^2}{2}\arcsin\frac{x}{a}$$

$$\int \frac{dx}{x^2\sqrt{a^2-x^2}} = -\frac{\sqrt{a^2-x^2}}{a^2x}$$

$$\int \frac{\sqrt{a^2-x^2}}{x^2}\,dx = -\frac{\sqrt{a^2-x^2}}{x} - \arcsin\frac{x}{a}$$

Function of $\sqrt{x^2-a^2}$

$$\int \sqrt{x^2-a^2}\,dx = \frac{1}{2}\left[x\sqrt{x^2-a^2} - a^2\ln\left(x+\sqrt{x^2-a^2}\right)\right]$$

$$\int \frac{dx}{\sqrt{x^2-a^2}} = \ln\left(x+\sqrt{x^2-a^2}\right)$$

$$\int \frac{dx}{x\sqrt{x^2-a^2}} = \frac{1}{a}\arccos\frac{a}{x}$$

$$\int \frac{\sqrt{x^2-a^2}}{x}\,dx = \sqrt{x^2-a^2} - a\arccos\frac{a}{x}$$

$$\int \frac{x\,dx}{\sqrt{x^2-a^2}} = \sqrt{x^2-a^2}$$

$$\int x\sqrt{x^2-a^2}\,dx = \frac{1}{3}(x^2-a^2)^{\frac{3}{2}}$$

$$\int \frac{x^2\,dx}{\sqrt{x^2-a^2}} = \frac{x}{2}\sqrt{x^2-a^2} + \frac{a^2}{2}\ln\left(x+\sqrt{x^2-a^2}\right)$$

Transcendental functions

$$\int xe^{ax}\,dx = \frac{e^{ax}}{a^2}(ax-1)$$

$$\int x^n e^{ax}\,dx = \frac{x^n e^{ax}}{a} - \frac{n}{a}\int x^{n-1}e^{ax}\,dx$$

$$\int \frac{e^{ax}}{x^n}\,dx = \frac{1}{n-1}\left[-\frac{e^{ax}}{x^{n-1}} + a\int \frac{e^{ax}}{x^{n-1}}\,dx\right]$$

$$\int a^{bx}\,dx = \frac{a^{bx}}{b\ln a}$$

$$\int x^n a^x \ dx = \frac{x^n a^x}{\ln a} - \frac{n}{\ln a}\int x^{n-1}a^x \ dx$$

$$\int \frac{dx}{a+e^x} = \ln\frac{e^x}{a+e^x}$$

$$\int \frac{dx}{a+be^{nx}} = \frac{1}{an}[nx - \ln(a+be^{nx})]$$

$$\int \frac{dx}{ae^{nx}+be^{-nx}} = \frac{1}{n\sqrt{ab}}\arctan\left(e^{nx}\sqrt{\frac{a}{b}}\right)$$

$$\int \ln x \ dx = x\ln x - x$$

$$\int \frac{(\ln x)^n}{x}\,dx = \frac{1}{n+1}(\ln x)^{n+1}$$

$$\int \sin^2 x \ dx = -\frac{1}{4}\sin 2x + \frac{1}{2}x = -\frac{1}{2}\sin x \cos x + \frac{1}{2}x$$

$$\int \cos^2 x \ dx = \frac{1}{4}\sin 2x + \frac{1}{2}x = \frac{1}{2}\sin x \cos x + \frac{1}{2}x$$

$$\int \sin nx \ dx = -\frac{\cos nx}{n}$$

$$\int \cos nx \ dx = \frac{\sin nx}{n}$$

$$\int \sin mx \cos nx \ dx = -\frac{\cos(m+n)x}{2(m+n)} - \frac{\cos(m-n)x}{2(m-n)} \quad (m^2 \neq n^2)$$

$$\int \sin mx \sin nx \ dx = \frac{\sin(m-n)x}{2(m-n)} - \frac{\sin(m+n)x}{2(m+n)} \quad (m^2 \neq n^2)$$

$$\int \cos mx \cos nx \ dx = \frac{\sin(m-n)x}{2(m-n)} + \frac{\sin(m+n)x}{2(m+n)} \quad (m^2 \neq n^2)$$

$$\int \frac{dx}{\sin x} = \ln\tan\frac{x}{2}$$

$$\int \frac{dx}{\cos x} = \ln\tan\left(\frac{\pi}{4}+\frac{x}{2}\right)$$

$$\int \frac{dx}{1+\cos x} = \tan\frac{x}{2}$$

$$\int \frac{dx}{1-\cos x} = -\cot\frac{x}{2}$$

$$\int \sin x \cos x \, dx = \frac{1}{2} \sin^2 x$$

$$\int \frac{dx}{\sin x \, \cos x} = \ln \tan x$$

$$\int \frac{dx}{a + b \cos x} = \frac{2}{\sqrt{a^2 - b^2}} \arctan\left(\sqrt{\frac{a-b}{a+b}} \tan \frac{x}{2}\right) \quad (a^2 > b^2)$$

$$= \frac{1}{\sqrt{b^2 - a^2}} \ln \frac{b + a \cos x + (\sin x)\sqrt{b^2 - a^2}}{a + b \cos x} \quad (a^2 < b^2)$$

$$\int \frac{\cos x \, dx}{a + b \cos x} = \frac{x}{b} - \frac{a}{b} \int \frac{dx}{a + b \cos x}$$

$$\int \frac{\sin x \, dx}{a + b \cos x} = -\frac{1}{b} \ln(a + b \cos x)$$

$$\int e^{ax} \sin bx \, dx = \frac{a \sin bx - b \cos bx}{a^2 + b^2} e^{ax}$$

$$\int e^{ax} \cos bx \, dx = \frac{a \cos bx + b \sin bx}{a^2 + b^2} e^{ax}$$

$$\int \arcsin x \, dx = x \arcsin x + \sqrt{1 - x^2}$$

$$\int \arccos x \, dx = x \arccos x - \sqrt{1 - x^2}$$

$$\int \arctan x \, dx = x \arctan x - \frac{1}{2} \ln(1 + x^2)$$

$$\int \operatorname{arccot} x \, dx = x \operatorname{arccot} x + \frac{1}{2} \ln(1 + x^2)$$

$$\int \frac{dx}{\sin^n x} = -\frac{\cos x}{(n-1)\sin^{n-1} x} + \frac{n-2}{n-1} \int \frac{dx}{\sin^{n-2} x}$$

$$\int \frac{dx}{\cos^n x} = \frac{\sin x}{(n-1)\cos^{n-1} x} + \frac{n-2}{n-1} \int \frac{dx}{\cos^{n-2} x}$$

Elementary differential equations

First order

Separation of variables:

$$\frac{dy}{dx} = f(x) \cdot g(y)$$

$$\int \frac{dy}{g(y)} = \text{const} = \int f(x)dx$$

Linear differential equation

$$\frac{dy}{dx} + f(x) \cdot y = g(x)$$

$$y = \frac{\left[\int (g(x)\exp\{\int f(x)dx\})dx + \text{const}\right]}{\exp\{\int f(x)dx\}}$$

Bernoulli (n ≠ 1)

$$\frac{dy}{dx} + f(x) \cdot y = g(x) \cdot y^n$$

$$y = \left[\frac{(1-n)\int g(x) \left[\exp\{(1-n)\int f(x)dx\}\right] dx + \text{const}}{\exp\{(1-n)\int f(x)dx\}}\right]^{1/1-n}$$

Bernoulli (n = 1)

$$\frac{dy}{dx} + f(x) \cdot y = g(x) \cdot y$$

$$y = \exp\left\{\int (g(x) - f(x))dx + \text{const}\right\}$$

Exact:

$$\frac{dy}{dx} = -\frac{f(x,y)}{g(x,y)} \text{ and } \frac{\partial f(x,y)}{\partial x} = \frac{\partial g(x,y)}{\partial y}$$

$$\int f(x,y)dx + \int \left(g(x,y) - \frac{\partial}{\partial y}\int f(x,y)dx\right)dy = \text{const}$$

First-order, linear homogeneous:

$$\frac{dy}{dx} + f(x)y = 0$$

$$y = \text{const} \exp\left\{-\int f(x)dx\right\}$$

First-order, linear inhomogeneous:

$$\frac{dy}{dx} + f(x) = g(x)$$

$$y = \frac{1}{\exp\int f(x)dx} \times \left\{\int f(x)g(x)\ dx + \text{const}\right\}$$

Second order

Bessel's differential equations:

$$x^2\frac{d^2y}{dx^2} + x\frac{dy}{dx} + (x^2 - n^2)y = 0$$

If order n is noninteger:

$$y = c_1 J_n(x) + c_2 J_{-n}(x)$$

If order n is integer:

$$y = c_1 J_n(x) + c_2 Y_n(x)$$

$$x^2\frac{d^2y}{dx^2} + ax\frac{dy}{dx} + (x^2 - n^2)y = 0$$

If order n is integer:

$$y = x^{\frac{1-a}{2}}\{c_1 J_n(ax) + c_2 Y_n(ax)\}$$

Modified Bessel's differential equations:

$$x^2\frac{d^2y}{dx^2} + x\frac{dy}{dx} - (x^2 + n^2)y = 0$$

$$y = c_1 I_n(x) + c_2 K_n(x)$$

where J is the Bessel functions of the first kind, Y is the Bessel function of the second kind, I is the modified Bessel function of the first kind, and K is the modified Bessel function of the second kind.

Second order, general

$$\frac{d^2y}{dx^2} + a\frac{dy}{dx} + by = R(x)$$

If u_1 and u_2 are solutions of $u^2 + au + b = 0$, three cases can be distinguished:
Case (1) u_1 and u_2 are real and different
Case (2) u_1 and u_2 are real and equal
Case (3) u_1 and u_2 are complex with $u_1 = p + qi$ and $u_2 = p - qi$

$$p = -\frac{1}{2}a \text{ and } q = \sqrt{b - \frac{1}{4}a^2}$$

Linear, homogeneous

$$\frac{d^2y}{dx^2} + a\frac{dy}{dx} + by = 0$$

Case (1) $y = c_1 e^{u_1 x} + c_2 e^{u_2 x}$
Case (2) $y = (c_1 + x\, c_2)\, e^{u_1 x}$
Case (3) $y = e^{px}(c_1 \cos qx + c_2 \sin qx)$

Linear, inhomogeneous

$$\frac{d^2y}{dx^2} + a\frac{dy}{dx} + by = R(x)$$

Case (1) $y = c_1 e^{u_1 x} + c_2 e^{u_2 x} + \dfrac{e^{u_1 x}}{u_1 - u_2} \displaystyle\int (e^{-u_1 x} R(x))\, dx$

$\qquad - \dfrac{e^{u_2 x}}{u_1 - u_2} \displaystyle\int (e^{-u_2 x} R(x))\, dx$

Case (2) $y = (c_1 + xc_2)e^{u_1 x} + x\, e^{u_1 x} \displaystyle\int (e^{-u_1 x} R(x))\, dx$

$\qquad - e^{u_1 x} \displaystyle\int (xe^{-u_1 x} R(x))\, dx$

Case (3) $y = e^{px} \left[(c_1 \cos qx + c_2 \sin qx) \right.$

$\qquad + \dfrac{\sin qx}{q} \displaystyle\int (e^{-p x} R(x) \cos qx)\, dx$

$\qquad \left. - \dfrac{\cos qx}{q} \displaystyle\int (e^{-p x} R(x) \sin qx)\, dx \right]$

Approximations for Bessel's functions

Notation	Abramowitz and Stegun	Jahnke, Emde, and Lösch	Whittaker and Watson
First kind	J_p	J_p	J_p
Second kind	Y_n	N_n	Y_n
Modified first kind	I_p	$i^{-p}J_p(ix)$	I_p
Modified second kind	K_n	$\frac{1}{2}\pi i^{n+1}H_n^1(ix)$	$\frac{K_n}{\cos n\pi}$

Approximations for small x (Abramowitz and Stegun notation)

$$J_p(x) \approx \frac{1}{2^p p!}x^p$$

$$J_{-p}(x) \approx \frac{2^p}{(-p)!}x^p$$

$$Y_0(x) \approx \frac{2}{\pi}\ln x$$

$$Y_n(x) \approx -\frac{2^n(n-1)!}{\pi}x^{-n},\ n \neq 0$$

$$I_p(x) \approx \frac{1}{2^p p!}x^p$$

$$I_{-p}(x) \approx \frac{2^p}{(-p)!}x^{-p}$$

$$K_0(x) \approx -\ln x$$

$$K_n(x) \approx 2^{n-1}(n-1)!x^{-n},\ n \neq 0$$

Approximations for large x (Abramowitz and Stegun notation)

$$J_p(x) \approx \sqrt{\frac{2}{\pi x}}\cos\left(x - \frac{\pi}{4} - \frac{p\pi}{2}\right)$$

$$Y_n(x) \approx \sqrt{\frac{2}{\pi x}}\sin\left(x - \frac{\pi}{4} - \frac{n\pi}{2}\right)$$

$$I_p(x) \approx \frac{e^x}{\sqrt{2\pi x}}$$

$$K_n(x) \approx \sqrt{\frac{\pi}{2x}} e^{-x}$$

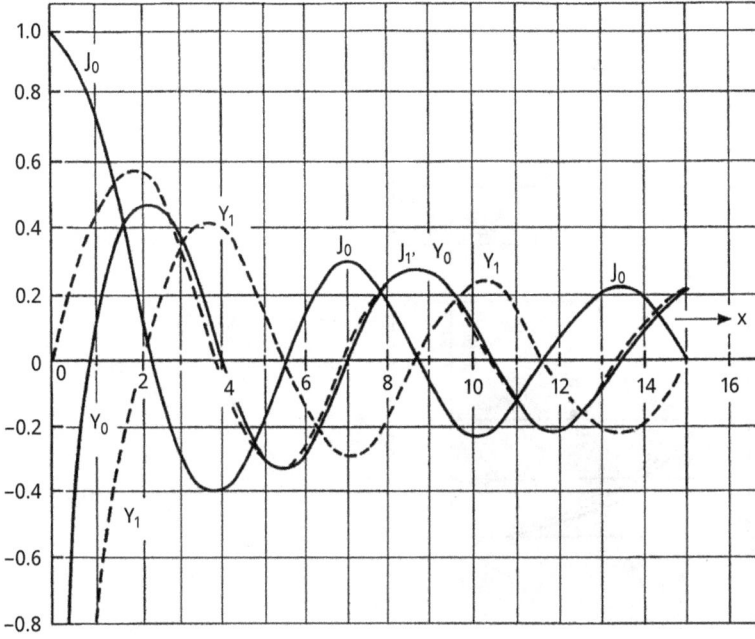

Fig 3.1: Bessel's functions.

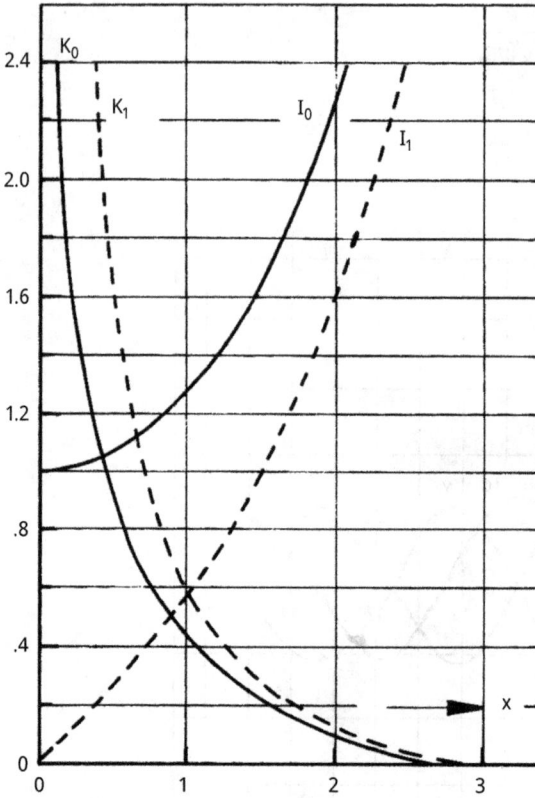

Fig. 3.2: Modified Bessel's functions.

The error function

The error function

$$\text{erf}(x) = \frac{2}{\sqrt{\pi}} \int_0^x e^{-t^2} dt$$

$$\text{erfc}(x) = 1 - \text{erf}(x)$$

$$\frac{d}{dx}\text{erf}(x) = \frac{2}{\sqrt{\pi}} e^{-x^2}$$

$$\int \text{erf}(x)\ dx = x\ \text{erf}(x) + \frac{e^{-x^2}}{\sqrt{\pi}} + \text{const}$$

Numerical approximation:

$$\text{erf}(x) \approx \frac{2}{\sqrt{\pi}} \text{sgn}(x)\sqrt{1 - e^{-x^2}}\left(\frac{\sqrt{\pi}}{2} + 0.155\ e^{-x^2} - 42.6 \times 10^{-3}\ e^{-2x^2}\right); \text{sgn}(x) = \frac{x}{|x|}$$

Expansion in series:

$$\text{erf}(x) = \frac{2}{\sqrt{\pi}}\left(x - \frac{x^3}{3 \cdot 1!} + \frac{x^5}{5 \cdot 2!} - \frac{x^7}{7 \cdot 3!} + \dots\right)$$

Properties:

$$\text{erf}(-x) = -\text{erf}(x)\quad \text{erf}(0) = 0\quad \lim_{x \to \infty} \text{erf}(x) = 1$$

Fig. 3.3: Error function.

x	**erf *x***	*x*	**erf *x***	*x*	**erf *x***
0	0.0000	1.00	0.8427	2.00	0.9953
0.05	0.0564	1.05	0.8624	2.05	0.9963
0.10	0.1125	1.10	0.8802	2.10	0.9970
0.15	0.1680	1.15	0.8961	2.15	0.9976
0.20	0.2227	1.20	0.9103	2.20	0.9981
0.25	0.2763	1.25	0.9229	2.25	0.9985
0.30	0.3286	1.30	0.9340	2.30	0.9989
0.35	0.3794	1.35	0.9438	2.35	0.9991
0.40	0.4284	1.40	0.9523	2.40	0.9993
0.45	0.4755	1.45	0.9597	2.45	0.9995
0.50	0.5205	1.50	0.9661	2.50	0.9996
0.55	0.5633	1.55	0.9716	3.00	0.99998
0.60	0.6039	1.60	0.9763	∞	1
0.65	0.6420	1.65	0.9804		
0.70	0.6778	1.70	0.9838		
0.75	0.7112	1.75	0.9867		
0.80	0.7421	1.80	0.9891		
0.85	0.7707	1.85	0.9911		
0.90	0.7969	1.90	0.9928		
0.95	0.8209	1.95	0.9942		

Laplace transforms

Notation

t	Variable in the real domain
$f(t)$	Function in the real domain
s	Variable in the complex domain (transformed t)
$F(s)$	Function in the complex domain (transformed $f(t)$)
a^-	$a^- \equiv \lim_{\epsilon \to 0} a - \epsilon$
	a and ϵ real; $\epsilon > 0$
a^+	$a^+ \equiv \lim_{\epsilon \to 0} a + \epsilon$
L	Transformation

Definition

$$L\{f(t)\} = F(s) = \int_{0^-}^{\infty} f(t)e^{-st}dt$$

with $f(t) \equiv 0$ for $t < 0$

Properties

Property	t-Domain	s-Domain
1. Linearity	$af_1(t) + bf_2(t)$	$aF_1(s) + bF_2(s)$
2. Conformity	$f(at)$	$\dfrac{1}{\lvert a \rvert}F\left(\dfrac{s}{a}\right)$
3. Translation		
a. Real	$f(t-\tau)$	$e^{-s\tau}F(s)$
b. Complex	$e^{-at}f(t)$	$F(s+a)$
(damping rule)		
4. Differentiation		
a. Real	$f^{(n)}(t) = \dfrac{d^n f(t)}{dt^n}$	$s^n F(s) - s^{n-1}f(0^-) - -$
(also valid if $f(t)$ is discontinuous at $t = 0$)	$n = 1, 2, \ldots$	$- - -f^{(n-1)}(0^-)$
b. Complex	$(-1)^n t^n f(t)$	$F^{(n)}(s) = \dfrac{d^n F(s)}{ds^n}$
	$n = 1, 2, \ldots$	
5. Integration		
a. Real	$\displaystyle\int_{0^-}^{t}\int_{0^-}^{t}\int_{0^-}^{t} \cdot \cdot \int_{0^-}^{t} f(\tau)d\tau \ldots d\tau$	$\dfrac{1}{s^n}F(s)$
(also valid if $f(t)$ is discontinuous at $t = 0$)	n times	

(continued)

Property	t-Domain	s-Domain
b. Complex	$\dfrac{1}{t}f(t)$	$\displaystyle\int_{s}^{\infty} F(u)\,du$ if $\displaystyle\lim_{t\to 0}\dfrac{1}{t}f(t)$ exists
6. Multiplication rule a. Real	$f_1(t)f_2(t)$	$\dfrac{1}{2\pi j}\displaystyle\int_{c-j\infty}^{c+j\infty} F_1(u)F_2(s^-)\,du$
b. Complex (convolution)	$f_1(t)\times f_2(t)$	$F_1(s)\ F_2(s)$
7. Autocorrelation	$\phi_{11}(t)=f(t)\times f(-t)$	$F(s)\ F(-s)$
8. Abel's theorems		
a. Begin value	$f(0^+)$	$\displaystyle\lim_{s\to\infty} s\,F(s)$
b. End value	$f(\infty)$	$\displaystyle\lim_{s\to 0} s\,F(s)$
9. Periodic functions	$f(t+T)=f(t)$	$\dfrac{\displaystyle\int_{0^-}^{T^-} f(\tau)e^{-st}\,d\tau}{1-e^{-sT}}$
10. Partial derivatives of $f(x,t)$ a. With respect to x	$\dfrac{\partial f(x,t)}{\partial x}$	$\dfrac{\partial F(x,s)}{\partial x}$
b. With respect to t	$\dfrac{\partial f(x,t)}{\partial t}$	$sF(x,s)-f(x,0^-)$
11. Inverse transform	$\dfrac{1}{2\pi j}\displaystyle\int_{c-j\infty}^{c+j\infty} e^{st}F(s)\,ds$	$F(s)$
12. Heaviside theorem	$\displaystyle\sum_{k=1}^{n}\dfrac{T(a_k)}{N(a_k)}e^{a_k t}$	$\dfrac{T(s)}{N(s)}$

$T(s)$ and $N(s)$ are polynomials in s. Power of T is smaller than the power of N.
$N(s)=(s-a_1)(s-a_2)\ldots(s-a_n)$, in which all a_n values are different.

Frequently used Laplace transforms

$F(s)$	$f(t)$, $t > 0$	
1	$\delta(t)$	Dirac pulse
e^{-sT}	$\delta(t-T)$	
$\dfrac{1}{s+a}$	e^{-at}	
$\dfrac{1}{(s+a)^2}$	$t\,e^{-at}$	
$\dfrac{1}{(s+a)(s+b)}$	$\dfrac{1}{b-a}\left(e^{-at}-e^{-bt}\right)$	
$\dfrac{\omega}{s^2+\omega^2}$	$\sin(\omega t)$	
$\dfrac{s}{s^2+\omega^2}$	$\cos(\omega t)$	
$\dfrac{1}{(s+a)^2+\omega^2}$	$\dfrac{1}{\omega}e^{-at}\sin(\omega t)$	
$\dfrac{1}{s^2+2\omega_n z+\omega_n^2}$	$\dfrac{1}{\beta}e^{-\omega_n z t}\sin(\beta t)$, $\beta=\omega_n\sqrt{1-z^2}$	
$\dfrac{1}{s}$	$U(t)$ of 1	Unit step
$\dfrac{1}{s}e^{-st}$	$U(t-T)$	
$\dfrac{1}{s}\left(1-e^{-sT}\right)$	$U(t)-U(t-T)$	
$\dfrac{1}{s(s+a)}$	$\dfrac{1}{a}\left(1-e^{-at}\right)$	
$\dfrac{1}{s(s+a)(s+b)}$	$\dfrac{1}{ab}\left(1-\dfrac{be^{-at}}{b-a}+\dfrac{ae^{-bt}}{b-a}\right)$	
$\dfrac{1}{s(s^2+\omega^2)}$	$\dfrac{1}{\omega^2}(1-\cos(\omega t))$	
$\dfrac{1}{s(s^2+2\omega_n z+\omega_n^2)}$	$\dfrac{1}{\omega_n^2}-\dfrac{1}{\omega_n\beta}e^{-\omega_n z t}\sin(\beta t+\gamma)$ $\beta=\omega_n\sqrt{1-z^2}$, $z=\cos\gamma$	
$\dfrac{1}{s(s+a)^2}$	$\dfrac{1}{a^2}\left(1-e^{-at}-ate^{-at}\right)$	
$\dfrac{1}{s^2}$	t	
$\dfrac{1}{s^2(s+a)}$	$\dfrac{1}{a^2}\left(at-1+e^{-at}\right)$	

Vector and tensors

Tensor $\underset{=}{\tau} = \begin{pmatrix} \tau_{xx} & \tau_{xy} & \tau_{xz} \\ \tau_{yx} & \tau_{yy} & \tau_{yz} \\ \tau_{zx} & \tau_{zy} & \tau_{zz} \end{pmatrix}$

Vector $\underline{v} = \begin{pmatrix} v_x & v_y & v_z \end{pmatrix}$

Scalar $s = s$

Vector mathematics

$$\underline{v} \pm \underline{w} = \begin{pmatrix} v_x \pm w_x, & v_y \pm w_y, v_z \pm w_z \end{pmatrix} = \sum_i \underline{\delta}_i (v_i + w_i)$$

$$s\,\underline{v} = \begin{pmatrix} sv_x, sv_y, sv_z \end{pmatrix} = \sum_i \underline{\delta}_i \,(sv_i)$$

$$\underline{v} \cdot \underline{w} = v_x w_x + v_y w_y + v_z w_z = \sum_i v_i w_i$$

$$\underline{v} \times \underline{w} = \begin{pmatrix} \delta_x & \delta_y & \delta_z \\ v_x & v_y & v_z \\ w_x & w_y & w_z \end{pmatrix} = \sum_i \sum_j \left[\underline{\delta}_i \times \underline{\delta}_j \right] v_i w_j$$

$$\underline{u} \cdot [\underline{v} \times \underline{w}] = \begin{pmatrix} u_x & u_y & u_z \\ v_x & v_y & v_z \\ w_x & w_y & w_z \end{pmatrix}$$

$$\underline{v}\,\underline{w} = \begin{pmatrix} v_x w_x & v_x w_y & v_x w_z \\ v_y w_x & v_y w_y & v_y w_z \\ v_z w_x & v_z w_y & v_z w_z \end{pmatrix} = \sum_i \sum_j \underline{\delta}_i \underline{\delta}_j v_i v_w$$

Tensor mathematics

$$\underset{=}{\sigma} \pm \underset{=}{\tau} = \begin{pmatrix} \sigma_{xx} \pm \tau_{xx} & \sigma_{xy} \pm \tau_{xy} & \sigma_{xz} \pm \tau_{xz} \\ \sigma_{yx} \pm \tau_{yx} & \sigma_{yy} \pm \tau_{yy} & \sigma_{yz} \pm \tau_{yz} \\ \sigma_{zx} \pm \tau_{zx} & \sigma_{zy} \pm \tau_{zy} & \sigma_{zz} \pm \tau_{zz} \end{pmatrix} = \sum_i \sum_j \underline{\delta}_i \underline{\delta}_j \left(\sigma_{ij} \pm \tau_{ij} \right)$$

$$s\,\underset{=}{\tau} = \begin{pmatrix} s\tau_{xx} & s\tau_{xy} & s\tau_{xz} \\ s\tau_{yx} & s\tau_{yy} & s\tau_{yz} \\ s\tau_{zx} & s\tau_{zy} & s\tau_{zz} \end{pmatrix} = \sum_i \sum_j \underline{\delta}_i \underline{\delta}_j \left(s\,\tau_{ij} \right)$$

$$\underline{\underline{\sigma}} : \underline{\underline{\tau}} = \sigma_{xx}\tau_{xx} + \sigma_{xy}\tau_{yx} + \sigma_{xz}\tau_{zx}$$

$$+ \sigma_{yx}\tau_{xy} + \sigma_{yy}\tau_{yy} + \sigma_{yz}\tau_{zy}$$

$$+ \sigma_{zx}\tau_{xz} + \sigma_{zy}\tau_{yz} + \sigma_{zz}\tau_{zz} = \sum_i \sum_j \sigma_{ij}\tau_{ij}$$

$$\underline{\underline{\sigma}} \cdot \underline{\underline{\tau}} = \begin{pmatrix} \sigma_{xx}\tau_{xx} + \sigma_{xy}\tau_{yx} + \sigma_{xz}\tau_{zx} & \sigma_{xx}\tau_{xy} + \sigma_{xy}\tau_{yy} + \sigma_{xz}\tau_{zy} & \sigma_{xx}\tau_{xz} + \sigma_{xy}\tau_{yz} + \sigma_{xz}\tau_{zz} \\ \sigma_{yx}\tau_{xx} + \sigma_{yy}\tau_{yx} + \sigma_{yz}\tau_{zx} & \sigma_{yx}\tau_{xy} + \sigma_{yy}\tau_{yy} + \sigma_{yz}\tau_{zy} & \sigma_{yx}\tau_{xz} + \sigma_{yy}\tau_{yz} + \sigma_{yz}\tau_{zz} \\ \sigma_{zx}\tau_{xx} + \sigma_{zy}\tau_{yx} + \sigma_{zz}\tau_{zx} & \sigma_{zx}\tau_{xy} + \sigma_{zy}\tau_{yy} + \sigma_{zz}\tau_{zy} & \sigma_{zx}\tau_{xz} + \sigma_{zy}\tau_{yz} + \sigma_{zz}\tau_{zz} \end{pmatrix}$$

$$= \sum_i \sum_j \underline{\delta}_i \underline{\delta}_j \left(\sum_l \sigma_{il}\tau_{lj} \right)$$

$$\underline{\underline{\tau}} \cdot \underline{v} = \left(\tau_{xx}v_x + \tau_{xy}v_y + \tau_{xz}v_z, \tau_{yx}v_x + \tau_{yy}v_y + \tau_{yz}v_z, \tau_{zx}v_x + \tau_{zy}v_y + \tau_{zz}v_z \right)$$

$$= \sum_i \underline{\delta}_i \left(\sum_j \tau_{ij}v_j \right)$$

Invariants of a tensor

First invariant

$$I_\tau = \sum_i \tau_{ii} = \text{tr } \underline{\underline{\tau}}$$

Second invariant

$$II_\tau = \sum_i \sum_j \tau_{ij}\tau_{ij} = \text{tr} \left(\underline{\underline{\tau}} \cdot \underline{\underline{\tau}} \right)$$

Third invariant

$$III_\tau = \sum_i \sum_j \sum_k \tau_{ij}\tau_{jk}\tau_{ki} = \text{tr} \left(\underline{\underline{\tau}} \cdot \left(\underline{\underline{\tau}} \cdot \underline{\underline{\tau}} \right) \right)$$

Also commonly used invariants

$$\overline{II}_\tau = \frac{1}{2} \left(I_\tau^2 - II_\tau \right)$$

$$\overline{III}_\tau = \frac{1}{6} \left(I_\tau^3 - 3\, I_\tau II_\tau + 2\, III_\tau \right) = \det \left| \underline{\underline{\tau}} \right|$$

Other vector and tensor operations

$$(\underline{\underline{\delta}} \cdot \underline{v}) = (\underline{v} \cdot \underline{\underline{\delta}}) = \underline{v}$$

$$(\underline{uv} \cdot \underline{w}) = \underline{u}(\underline{v} \cdot \underline{w})$$

$$(\underline{w} \cdot \underline{uv}) = (\underline{w} \cdot \underline{u})\underline{v}$$

$$\underline{uv} : \underline{wz} = (\underline{uw} : \underline{vz}) = (\underline{u} \cdot \underline{z})(\underline{v} \cdot \underline{w})$$

$$\underline{\underline{\tau}} : \underline{uv} = \left(\underline{\underline{\tau}} \cdot \underline{u} \right) \cdot \underline{v}$$

$$\underline{uv} : \underline{\underline{\tau}} = u \cdot \left(\underline{v} \cdot \underline{\underline{\tau}} \right)$$

Differentiation in vector and tensor notation

Notation

r, s: scalar
$\underline{v}$, $\underline{w}$: vector
$\underline{\underline{\tau}}$: tensor
$\underline{\delta}$: unit vector
$\underline{\underline{\delta}}$: unit tensor
$\underline{\delta}_x$: unit vector in the x direction
For summation, the convention $x_1 = x$, $x_2 = y$, $x_3 = z$ is used.

Nabla operator

Cartesian coordinate system

$$\nabla = \underline{\delta}_x \frac{\partial}{\partial x} + \underline{\delta}_y \frac{\partial}{\partial y} + \underline{\delta}_z \frac{\partial}{\partial z} = \sum_i \underline{\delta}_i \frac{\partial}{\partial x_i}$$

Cylindrical coordinate system

$$\nabla = \underline{\delta}_r \frac{\partial}{\partial r} + \underline{\delta}_\vartheta \frac{1}{r} \frac{\partial}{\partial \vartheta} + \underline{\delta}_z \frac{\partial}{\partial z}$$

Spherical coordinate system

$$\nabla = \underline{\delta}_r \frac{\partial}{\partial r} + \underline{\delta}_\vartheta \frac{1}{r} \frac{\partial}{\partial \vartheta} + \underline{\delta}_\varphi \frac{1}{r \sin \vartheta} \frac{\partial}{\partial \varphi}$$

Nabla operations in a Cartesian coordinate system

Gradient of a scalar field

$$\nabla s = \underline{\delta}_x \frac{\partial s}{\partial x} + \underline{\delta}_y \frac{\partial s}{\partial y} + \underline{\delta}_z \frac{\partial s}{\partial z} = \sum_i \underline{\delta}_i \frac{\partial s}{\partial x_i}$$

Divergence of a vector field

$$\nabla \cdot \underline{v} = \frac{\partial v_x}{\partial x} + \frac{\partial v_y}{\partial y} + \frac{\partial v_z}{\partial z} = \sum_i \frac{\partial v_i}{\partial x_i}$$

Curl of a vector field

$$\nabla \times \underline{v} = \underline{\delta}_x \left\{ \frac{\partial v_z}{\partial y} - \frac{\partial v_y}{\partial z} \right\} + \underline{\delta}_y \left\{ \frac{\partial v_x}{\partial z} - \frac{\partial v_z}{\partial x} \right\} + \underline{\delta}_z \left\{ \frac{\partial v_y}{\partial x} - \frac{\partial v_x}{\partial y} \right\} = \sum_i \sum_j \left[\underline{\delta}_i \times \underline{\delta}_j \right] \frac{\partial v_j}{\partial x_i}$$

Laplacian of a scalar field

$$\nabla \cdot \nabla s = \frac{\partial^2 s}{\partial x^2} + \frac{\partial^2 s}{\partial y^2} + \frac{\partial^2 s}{\partial z^2} = \sum_i \frac{\partial^2 s}{\partial x_i^2}$$

Laplacian of a vector field

$$\nabla \cdot \nabla \underline{v} = \underline{\delta}_x \left\{ \frac{\partial^2 v_x}{\partial x^2} + \frac{\partial^2 v_x}{\partial y^2} + \frac{\partial^2 v_x}{\partial z^2} \right\} + \underline{\delta}_y \left\{ \frac{\partial^2 v_y}{\partial x^2} + \frac{\partial^2 v_y}{\partial y^2} + \frac{\partial^2 v_y}{\partial z^2} \right\} +$$

$$+ \underline{\delta}_z \left\{ \frac{\partial^2 v_z}{\partial x^2} + \frac{\partial^2 v_z}{\partial y^2} + \frac{\partial^2 v_z}{\partial z^2} \right\} = \sum_k \underline{\delta}_k \left\{ \sum_i \frac{\partial^2 v_k}{\partial x_i^2} \right\}$$

Nabla identities

$$\nabla \, r \, s = r \nabla s + s \nabla r$$

$$\nabla \cdot s\underline{v} = (\nabla s \cdot \underline{v}) + s(\nabla \cdot \underline{v})$$

$$\nabla \cdot (\underline{v} \times \underline{w}) = \underline{w} \cdot (\nabla \times \underline{v}) - \underline{v} \cdot (\nabla \cdot \underline{w})$$

$$\nabla \times s\underline{v} = (\nabla s \times \underline{v}) + s(\nabla \times \underline{v})$$

$$\nabla \cdot \nabla \underline{v} = \nabla(\nabla \cdot \underline{v}) - \nabla \times (\nabla \times \underline{v})$$

$$\underline{v} \cdot \nabla \underline{v} = 1/2 \nabla (\underline{v} \cdot \underline{v}) - \underline{v} \times (\nabla \times \underline{v})$$

$$\nabla \cdot \underline{v}w = \underline{v} \cdot \nabla \underline{w} + w(\nabla \cdot v)$$

$$\nabla \cdot s \, \underline{\underline{\tau}} = \nabla s \cdot \underline{\underline{\tau}} + s \left(\nabla \cdot \underline{\underline{\tau}} \right)$$

Scalar derivatives with respect to time

Partial derivative

$$\frac{\partial c}{\partial t}$$

Total derivative

$$\frac{dc}{dt} = \frac{\partial c}{\partial t} + \frac{\partial c}{\partial x}\frac{\partial x}{\partial t} + \frac{\partial c}{\partial y}\frac{\partial y}{\partial t} + \frac{\partial c}{\partial z}\frac{\partial z}{\partial t} = \frac{\partial c}{\partial t} + \sum_i \frac{\partial c}{\partial x_i}\frac{\partial x_i}{\partial t}$$

Substantial derivative

$$\frac{Dc}{Dt} = \frac{\partial c}{\partial t} + v_x\frac{\partial c}{\partial x} + v_y\frac{\partial c}{\partial y} + v_z\frac{\partial c}{\partial z} = \frac{\partial c}{\partial t} + \sum_i v_i\frac{\partial c}{\partial x_i}$$

Tensorial derivatives with respect to time

Codeformational or Oldroyd derivative

$$\frac{\delta \tau_{ij}}{\delta t} = \frac{\partial \tau_{ij}}{\partial t} + v_x\frac{\partial \tau_{ij}}{\partial x} + v_y\frac{\partial \tau_{ij}}{\partial y} + v_z\frac{\partial \tau_{ij}}{\partial z} - \tau_{xj}\frac{\partial v_i}{\partial x} - \tau_{yj}\frac{\partial v_i}{\partial y} - \tau_{zj}\frac{\partial v_i}{\partial z} - \tau_{ix}\frac{\partial v_j}{\partial x} - \tau_{iy}\frac{\partial v_j}{\partial y} - \tau_{iz}\frac{\partial v_j}{\partial z}$$

$$= \frac{\partial \tau_{ij}}{\partial t} + \sum_k \left\{ v_k\frac{\partial \tau_{ij}}{\partial x_k} - \tau_{kj}\frac{\partial v_i}{\partial x_k} - \tau_{ik}\frac{\partial v_j}{\partial x_k} \right\} \quad (i,j = x,y,z)$$

corotational or Jaumann derivative

$$\frac{\mathcal{D}\tau_{ij}}{\mathcal{D}t} = \frac{\partial \tau_{ij}}{\partial t} + \sum_k v_k\frac{\partial \tau_{ij}}{\partial x_k} - \frac{1}{2}\sum_m \left(\omega_{jm}\tau_{mi} + \omega_{im}\tau_{mj}\right) \text{ with } \omega_{ij} = \frac{\partial v_i}{\partial x_j} - \frac{\partial v_j}{\partial x_i}$$

Linear regression

"Curve fit" of linear, exponential, and power functions

Average values

$$\bar{x} = \left(\sum x\right)/n$$

$$\bar{y} = \left(\sum y\right)/n$$

Standard deviations

$$S_x = \sqrt{\frac{\sum x^2 - \left(\sum x\right)^2/n}{n-1}}$$

$$S_y = \sqrt{\frac{\sum y^2 - \left(\sum y\right)^2/n}{n-1}}$$

$y = a + bx$

$$b = \frac{n\sum xy - \sum x \sum y}{n\sum x^2 - \left(\sum x\right)^2}$$

$$a = \frac{\sum y - b\sum x}{n}$$

$$r = b\sqrt{\frac{n\sum x^2 - \left(\sum x\right)^2}{n\sum y^2 - \left(\sum y\right)^2}} = b\frac{S_x}{S_y}$$

$y = a \ \exp(bx)$

$$b = \frac{n\sum(x\ln y) - \sum x\sum \ln y}{n\sum x^2 - \left(\sum x\right)^2}$$

$$a = \exp\frac{\sum \ln y - b\sum x}{n}$$

$$r = b\sqrt{\frac{n\sum x^2 - \left(\sum x\right)^2}{n\sum (\ln y)^2 - \left(\sum \ln y\right)^2}}$$

$y = ax^b$

$$b = \frac{n \sum (\ln x \ln y) - \sum \ln x \sum \ln y}{n \sum (\ln x)^2 - (\sum \ln x)^2}$$

$$a = \exp \frac{\sum \ln y - b \sum \ln x}{n}$$

$$r = b \sqrt{\frac{n \sum (\ln x)^2 - (\sum \ln x)^2}{n \sum (\ln y)^2 - (\sum \ln y)^2}}$$

Chapter 4
Transport phenomena

Concentration notation

Standard quantities

Density	$\rho = \rho_i + \rho_j$	[kg/m^3]
Molar concentration	$c = c_i + c_j$	[mol/m^3]

Derived quantities

Mass fraction	$w_i = \rho_i/\rho$	[–]
Mole fraction	$x_i = c_i/c$	[–]
Molar mass	$M_i = \rho_i/c_i$	[kg/mol]
Number averaged molar mass	$M = \rho/c$	[kg/mol]

Relations

$$x_i + x_j = 1$$

$$w_i + w_j = 1$$

$$x_i M_i + x_j M_j = M$$

$$\frac{w_i}{M_i} + \frac{w_j}{M_j} = \frac{1}{M}$$

$$x_i = \frac{w_i M_j}{w_i M_j + w_j M_i}$$

$$w_i = \frac{x_i M_i}{x_i M_i + x_j M_j}$$

$$dx_i = \frac{M_i M_j}{\left(w_i M_j + w_j M_i\right)^2} \, dw_i$$

$$dw_i = \frac{M_i M_j}{\left(x_i M_i + x_j M_j\right)^2} \, dx_i$$

https://doi.org/10.1515/9783111385341-004

Microscopic balances in general form

Notation
$x, y, z, r, \vartheta, \varphi$: coordinates $\qquad$ C_V: specific heat at constant volume
ρ: density $\qquad$ T: temperature
v: velocity $\qquad$ Q'': heat flux
t: time $\qquad$ C: molar concentration
τ: shear stress $\qquad$ R: molar production
g: gravitational acceleration $\qquad$ N: molar flux
p: pressure

Equation of continuity
Cartesian coordinate system (x, y, z)

$$\frac{\partial \rho}{\partial t} + \frac{\partial}{\partial x}(\rho v_x) + \frac{\partial}{\partial y}(\rho v_y) + \frac{\partial}{\partial z}(\rho v_z) = 0$$

Cylindrical coordinate system (r, ϑ, z)

$$\frac{\partial \rho}{\partial t} + \frac{1}{r}\frac{\partial}{\partial r}(\rho r v_r) + \frac{1}{r}\frac{\partial}{\partial \vartheta}(\rho v_\vartheta) + \frac{\partial}{\partial z}(\rho v_z) = 0$$

Spherical coordinate system (r, ϑ, φ)

$$\frac{\partial \rho}{\partial t} + \frac{1}{r^2}\frac{\partial}{\partial r}(\rho r^2 v_r) + \frac{1}{r \sin \vartheta}\frac{\partial}{\partial \vartheta}(\rho v_\vartheta \sin \vartheta) + \frac{1}{r \sin \vartheta}\frac{\partial}{\partial \varphi}(\rho v_\varphi) = 0$$

Mass balance for component A
Cartesian coordinate system (x, y, z)

$$\frac{\partial c_A}{\partial t} + \left(\frac{\partial N_{Ax}}{\partial x} + \frac{\partial N_{Ay}}{\partial y} + \frac{\partial N_{Az}}{\partial z}\right) = R_A$$

Cylindrical coordinate system (r, ϑ, z)

$$\frac{\partial c_A}{\partial t} + \left(\frac{1}{r}\frac{\partial}{\partial r}(rN_{Ar}) + \frac{1}{r}\frac{\partial N_{A\vartheta}}{\partial \vartheta} + \frac{\partial N_{Az}}{\partial z}\right) = R_A$$

Spherical coordinate system (r, ϑ, φ)

$$\frac{\partial c_A}{\partial t} + \left(\frac{1}{r^2}\frac{\partial}{\partial r}(r^2 N_{Ar}) + \frac{1}{r \sin \vartheta}\frac{\partial}{\partial \vartheta}(N_{A\vartheta} \sin \vartheta) + \frac{1}{r \sin \vartheta}\frac{\partial N_{A\varphi}}{\partial \varphi}\right) = R_A$$

Equations of motion (Navier-Stokes equations)

Cartesian coordinate system (x, y, z)

$$\rho\left(\frac{\partial v_x}{\partial t} + v_x\frac{\partial}{\partial x}v_x + v_y\frac{\partial}{\partial y}v_x + v_z\frac{\partial}{\partial z}v_x\right) = -\left[\frac{\partial}{\partial x}\tau_{xx} + \frac{\partial}{\partial y}\tau_{yx} + \frac{\partial}{\partial z}\tau_{zx}\right] - \frac{\partial p}{\partial x} + \rho g_x$$

$$\rho\left(\frac{\partial v_y}{\partial t} + v_x\frac{\partial}{\partial x}v_y + v_y\frac{\partial}{\partial y}v_y + v_z\frac{\partial}{\partial z}v_y\right) = -\left[\frac{\partial}{\partial x}\tau_{xy} + \frac{\partial}{\partial y}\tau_{yy} + \frac{\partial}{\partial z}\tau_{zy}\right] - \frac{\partial p}{\partial y} + \rho g_y$$

$$\rho\left(\frac{\partial v_z}{\partial t} + v_x\frac{\partial}{\partial x}v_z + v_y\frac{\partial}{\partial y}v_z + v_z\frac{\partial}{\partial z}v_z\right) = -\left[\frac{\partial}{\partial x}\tau_{xz} + \frac{\partial}{\partial y}\tau_{yz} + \frac{\partial}{\partial z}\tau_{zz}\right] - \frac{\partial p}{\partial z} + \rho g_z$$

Cylindrical coordinate system (r, ϑ, z)

$$\rho\left(\frac{\partial v_r}{\partial t} + v_r\frac{\partial v_r}{\partial r} + \frac{v_\vartheta}{r}\frac{\partial v_r}{\partial \vartheta} - \frac{v_\vartheta^2}{r} + v_z\frac{\partial v_r}{\partial z}\right) = -\left[\frac{1}{r}\frac{\partial}{\partial r}(r\tau_{rr}) + \frac{1}{r}\frac{\partial}{\partial \vartheta}\tau_{\vartheta r} + \frac{\partial}{\partial z}\tau_{zr} - \frac{\tau_{\vartheta\vartheta}}{r}\right]$$
$$- \frac{\partial p}{\partial r} + \rho g_r$$

$$\rho\left(\frac{\partial v_\vartheta}{\partial t} + v_r\frac{\partial v_\vartheta}{\partial r} + \frac{v_\vartheta}{r}\frac{\partial v_\vartheta}{\partial \vartheta} + \frac{v_r v_\vartheta}{r} + v_z\frac{\partial v_\vartheta}{\partial z}\right) = -\left[\frac{1}{r^2}\frac{\partial}{\partial r}(r^2\tau_{r\vartheta}) + \frac{1}{r}\frac{\partial}{\partial \vartheta}\tau_{\vartheta\vartheta} + \frac{\partial}{\partial z}\tau_{z\vartheta} + \frac{\tau_{\vartheta r} - \tau_{r\vartheta}}{r}\right]$$
$$- \frac{1}{r}\frac{\partial p}{\partial \vartheta} + \rho g_\vartheta$$

$$\rho\left(\frac{\partial v_z}{\partial t} + v_r\frac{\partial v_z}{\partial r} + \frac{v_\vartheta}{r}\frac{\partial v_z}{\partial \vartheta} + v_z\frac{\partial v_z}{\partial z}\right) = -\left[\frac{1}{r}\frac{\partial}{\partial r}(r\tau_{rz}) + \frac{1}{r}\frac{\partial}{\partial \vartheta}\tau_{\vartheta z} + \frac{\partial}{\partial z}\tau_{zz}\right] - \frac{\partial p}{\partial z} + \rho g_z$$

Spherical coordinate system (r, ϑ, φ)

$$\rho\left(\frac{\partial v_r}{\partial t} + v_r\frac{\partial v_r}{\partial r} + \frac{v_\vartheta}{r}\frac{\partial v_r}{\partial \vartheta} + \frac{v_\varphi}{r\sin\vartheta}\frac{\partial v_r}{\partial \varphi} - \frac{v_\vartheta^2 + v_\varphi^2}{r}\right)$$
$$= -\left[\frac{1}{r^2}\frac{\partial}{\partial r}(r^2\tau_{rr}) + \frac{1}{r\sin\vartheta}\frac{\partial}{\partial \vartheta}(\tau_{\vartheta r}\sin\vartheta) + \frac{1}{r\sin\vartheta}\frac{\partial}{\partial \varphi}\tau_{\varphi r} - \frac{\tau_{\vartheta\vartheta} + \tau_{\varphi\varphi}}{r}\right] - \frac{\partial p}{\partial r} + \rho g_r$$

$$\rho\left(\frac{\partial v_\vartheta}{\partial t} + v_r\frac{\partial v_\vartheta}{\partial r} + \frac{v_\vartheta}{r}\frac{\partial v_\vartheta}{\partial \vartheta} + \frac{v_\varphi}{r\sin\vartheta}\frac{\partial v_\vartheta}{\partial \varphi} + \frac{v_r v_\vartheta}{r} - \frac{v_\varphi^2\cot\vartheta}{r}\right)$$
$$= -\left[\frac{1}{r^3}\frac{\partial}{\partial r}(r^3\tau_{r\vartheta}) + \frac{1}{r\sin\vartheta}\frac{\partial}{\partial \vartheta}(\tau_{\vartheta\vartheta}\sin\vartheta) + \frac{1}{r\sin\vartheta}\frac{\partial}{\partial \varphi}\tau_{\varphi\vartheta} + \frac{(\tau_{\vartheta r} - \tau_{r\vartheta}) - \tau_{\varphi\varphi}\cot\vartheta}{r}\right] - \frac{1}{r}\frac{\partial p}{\partial \vartheta} + \rho g_\vartheta$$

$$\rho\left(\frac{\partial v_\varphi}{\partial t} + v_r\frac{\partial v_\varphi}{\partial r} + \frac{v_\vartheta}{r}\frac{\partial v_\varphi}{\partial \vartheta} + \frac{v_\varphi}{r\sin\vartheta}\frac{\partial v_\varphi}{\partial \varphi} + \frac{v_\varphi v_r}{r} + \frac{v_\vartheta v_\varphi}{r}\cot\vartheta\right)$$

$$= -\left[\frac{1}{r^3}\frac{\partial}{\partial r}\left(r^3\tau_{r\varphi}\right) + \frac{1}{r\sin\vartheta}\frac{\partial}{\partial\vartheta}\left(\tau_{\vartheta\varphi}\sin\vartheta\right) + \frac{1}{r\sin\vartheta}\frac{\partial}{\partial\varphi}\tau_{\varphi\varphi} + \frac{\left(\tau_{\varphi r} - \tau_{r\varphi}\right) + \tau_{\varphi\vartheta}\cot\vartheta}{r}\right]$$

$$- \frac{1}{r\sin\vartheta}\frac{\partial p}{\partial\varphi} + \rho g_\varphi$$

Energy balance

Cartesian coordinate system (x, y, z)

$$\rho C_v\left(\frac{\partial T}{\partial t} + v_x\frac{\partial T}{\partial x} + v_y\frac{\partial T}{\partial y} + v_z\frac{\partial T}{\partial z}\right)$$

$$= -\left[\frac{\partial Q''_x}{\partial x} + \frac{\partial Q''_y}{\partial y} + \frac{\partial Q''_z}{\partial z}\right] - T\left(\frac{\partial p}{\partial T}\right)_\rho\left(\frac{\partial v_x}{\partial x} + \frac{\partial v_y}{\partial y} + \frac{\partial v_z}{\partial z}\right)$$

$$- \left\{\tau_{xx}\frac{\partial v_x}{\partial x} + \tau_{yy}\frac{\partial v_y}{\partial y} + \tau_{zz}\frac{\partial v_z}{\partial z}\right\}$$

$$- \left\{\tau_{xy}\left(\frac{\partial v_x}{\partial y} + \frac{\partial v_y}{\partial x}\right) + \tau_{xz}\left(\frac{\partial v_x}{\partial z} + \frac{\partial v_z}{\partial x}\right) + \tau_{yz}\left(\frac{\partial v_y}{\partial z} + \frac{\partial v_z}{\partial y}\right)\right\}$$

Cylindrical coordinate system (r, ϑ, z)

$$\rho C_v\left(\frac{\partial T}{\partial t} + v_r\frac{\partial T}{\partial r} + \frac{v_\vartheta}{r}\frac{\partial T}{\partial\vartheta} + v_z\frac{\partial T}{\partial z}\right)$$

$$= -\left[\frac{1}{r}\frac{\partial}{\partial r}\left(rQ''_r\right) + \frac{1}{r}\frac{\partial Q''_\vartheta}{\partial\vartheta} + \frac{\partial Q''_z}{\partial z}\right] - T\left(\frac{\partial p}{\partial T}\right)_\rho\left(\frac{1}{r}\frac{\partial}{\partial r}\left(rv_r\right) + \frac{1}{r}\frac{\partial v_\vartheta}{\partial\vartheta} + \frac{\partial v_z}{\partial z}\right)$$

$$- \left\{\tau_{rr}\frac{\partial v_r}{\partial r} + \tau_{\vartheta\vartheta}\frac{1}{r}\left(\frac{\partial v_\vartheta}{\partial\vartheta} + v_r\right) + \tau_{zz}\frac{\partial v_z}{\partial z}\right\}$$

$$- \left\{\tau_{r\vartheta}\left[r\frac{\partial}{\partial r}\left(\frac{v_\vartheta}{r}\right) + \frac{1}{r}\frac{\partial v_r}{\partial\vartheta}\right] + \tau_{rz}\left(\frac{\partial v_z}{\partial r} + \frac{\partial v_r}{\partial z}\right) + \tau_{\vartheta z}\left(\frac{1}{r}\frac{\partial v_z}{\partial\vartheta} + \frac{\partial v_\vartheta}{\partial z}\right)\right\}$$

Spherical coordinate system (r, ϑ, φ)

$$\rho C_v \left(\frac{\partial T}{\partial t} + v_r \frac{\partial T}{\partial r} + \frac{v_\vartheta}{r} \frac{\partial T}{\partial \vartheta} + \frac{v_\varphi}{r \sin \vartheta} \frac{\partial T}{\partial \varphi} \right)$$

$$= - \left[\frac{1}{r^2} \frac{\partial}{\partial r} (r^2 Q_r'') + \frac{1}{r \sin \vartheta} \frac{\partial}{\partial \vartheta} (Q_\vartheta'' \sin \vartheta + \frac{1}{r \sin \vartheta} \frac{\partial Q_\varphi''}{\partial \varphi} \right]$$

$$- T \left(\frac{\partial p}{\partial T} \right)_\rho \left(\frac{1}{r^2} \frac{\partial}{\partial r} (r^2 v_r) + \frac{1}{r \sin \vartheta} \frac{\partial}{\partial \vartheta} (v_\vartheta \sin \vartheta) + \frac{1}{r \sin \vartheta} \frac{\partial v_\varphi}{\partial \varphi} \right)$$

$$- \left\{ \tau_{rr} \frac{\partial v_r}{\partial r} + \tau_{\vartheta\vartheta} \left(\frac{1}{r} \frac{\partial v_\vartheta}{\partial \vartheta} + \frac{v_r}{r} \right) + \tau_{\varphi\varphi} \left(\frac{1}{r \sin \vartheta} \frac{\partial v_\varphi}{\partial \varphi} + \frac{v_r}{r} + \frac{v_\vartheta \cot \vartheta}{r} \right) \right\}$$

$$- \left\{ \tau_{r\vartheta} \left(\frac{\partial v_\vartheta}{\partial r} + \frac{1}{r} \frac{\partial v_r}{\partial \vartheta} - \frac{v_\vartheta}{r} \right) + \tau_{r\varphi} \left(\frac{\partial v_\varphi}{\partial r} + \frac{1}{r \sin \vartheta} \frac{\partial v_r}{\partial \varphi} - \frac{v_\varphi}{r} \right) \right.$$

$$\left. + \tau_{\vartheta\varphi} \left(\frac{1}{r} \frac{\partial v_\varphi}{\partial \vartheta} + \frac{1}{r \sin \vartheta} \frac{\partial v_\vartheta}{\partial \varphi} - \frac{\cot \vartheta}{r} v_\varphi \right) \right\}$$

Continuity equation in tensorial form
Compressible

$$\frac{\partial}{\partial t} \rho = -(\nabla \cdot \rho \underline{v})$$

Incompressible

$$\nabla \cdot \underline{v} = 0$$

Equation of motion in tensorial form

$$\frac{\partial}{\partial t} (\rho \underline{v}) = -(\nabla \cdot \rho \underline{v} \, \underline{v}) - \left[\nabla \cdot \left(p \underline{\underline{\delta}} + \underline{\underline{\tau}} \right) \right] + \rho \underline{g}$$

Bernoulli's equations

Notation

ϕ_m	: mass flow rate	R	: gas constant
ϕ_A	: total energy added	M	: molecular weight
A_{wr}	:energy dissipation per unit mass	κ	: C_v/C_p
K_w	: friction loss factor	D_h	: hydraulic diameter
f	: friction coefficient		

General Bernoulli's equation

$$0 = -\left\{ \int_1^2 \frac{dp}{\rho} + g(h_2 - h_1) + \tfrac{1}{2}(<v_2>^2 - <v_1>)^2 \right\} \phi_{\mathrm{m}} + \phi_{\mathrm{A}} - A_{\mathrm{wr}} \phi_{\mathrm{m}}$$

$$\text{with } A_{\mathrm{wr}} = \sum_i \left(4f \frac{1}{2} \langle v \rangle^2 \frac{L}{D_{\mathrm{h}}} \right)_i + \sum_j \left(K_{\mathrm{w}} \frac{1}{2} <v>^2 \right)_j$$

Frictionless flow of an incompressible medium with $\phi_A = 0$

$$\frac{p}{\rho} + gh + \tfrac{1}{2} <v>^2 = \text{constant along a stream line}$$

Isothermal flow of an ideal gas

$$\int_1^2 \frac{dp}{\rho} = \frac{RT}{M} \ln \frac{p_2}{p_1}$$

Adiabatic flow of an ideal gas

$$\int_1^2 \frac{dp}{\rho} = \frac{p_1}{\rho_1} \frac{\kappa}{\kappa - 1} \left\{ \left(\frac{p_2}{p_1} \right)^{\frac{\kappa-1}{\kappa}} - 1 \right\}$$

Microscopic balances for idealized materials

Newtonian medium with constant viscosity, specific heat, density, heat conductivity, diffusivity, and density

Notation

$x, y, z, r, \vartheta, \varphi$: coordinates g: gravitational acceleration
ρ: density C_p: specific heat at constant pressure
v: velocity λ: heat conductivity
η: dynamic viscosity C: molar concentration
p: pressure $\mathcal{D}$: diffusion coefficient
t: time R: molar production
T: temperature q: heat production per volume unit

Continuity equations (total mass balances)

$$\nabla \cdot \boldsymbol{v} = 0$$

Rectangular coordinate system

$$\frac{\partial v_x}{\partial x} + \frac{\partial v_y}{\partial y} + \frac{\partial v_z}{\partial z} = 0$$

Cylindrical coordinate system

$$\frac{1}{r}\frac{\partial (rv_r)}{\partial r} + \frac{1}{r}\frac{\partial v_\theta}{\partial \theta} + r\frac{\partial v_z}{\partial z} = 0$$

Spherical coordinate system

$$\frac{1}{r^2}\frac{\partial}{\partial r}\left(r^2 v_r\right) + \frac{1}{r\sin\theta}\frac{\partial}{\partial \theta}(v_\theta \sin\theta) + \frac{1}{r\sin\theta}\frac{\partial v_\varphi}{\partial \varphi} = 0$$

Navier-Stokes equations (momentum balances)

$$\rho\frac{D\boldsymbol{v}}{Dt} = \mu\nabla^2\boldsymbol{v} - \nabla p + \rho\boldsymbol{g}$$

Rectangular coordinate system
x-Coordinate

$$\rho\left(\frac{\partial v_x}{\partial t} + v_x\frac{\partial v_x}{\partial x} + v_y\frac{\partial v_x}{\partial y} + v_z\frac{\partial v_x}{\partial z}\right) = \mu\left(\frac{\partial^2 v_x}{\partial x^2} + \frac{\partial^2 v_x}{\partial y^2} + \frac{\partial^2 v_x}{\partial z^2}\right) - \frac{\partial p}{\partial x} + \rho g_x$$

y-Coordinate

$$\rho\left(\frac{\partial v_y}{\partial t} + v_x\frac{\partial v_y}{\partial x} + v_y\frac{\partial v_y}{\partial y} + v_z\frac{\partial v_y}{\partial z}\right) = \mu\left(\frac{\partial^2 v_y}{\partial x^2} + \frac{\partial^2 v_y}{\partial y^2} + \frac{\partial^2 v_y}{\partial z^2}\right) - \frac{\partial p}{\partial y} + \rho g_y$$

z-Coordinate

$$\rho\left(\frac{\partial v_z}{\partial t} + v_x\frac{\partial v_z}{\partial x} + v_y\frac{\partial v_z}{\partial y} + v_z\frac{\partial v_z}{\partial z}\right) = \mu\left(\frac{\partial^2 v_z}{\partial x^2} + \frac{\partial^2 v_z}{\partial y^2} + \frac{\partial^2 v_z}{\partial z^2}\right) - \frac{\partial p}{\partial z} + \rho g_z$$

Cylindrical coordinate system
r-Coordinate (radial)

$$\rho\left(\frac{\partial v_r}{\partial t} + v_r\frac{\partial v_r}{\partial r} + \frac{v_\theta}{r}\frac{\partial v_r}{\partial \theta} - \frac{v_\theta^2}{r} + v_z\frac{\partial v_r}{\partial z}\right)$$

$$= \mu\left(\frac{\partial}{\partial r}\left(\frac{1}{r}\frac{\partial}{\partial r}(rv_r)\right) + \frac{1}{r^2}\frac{\partial^2 v_r}{\partial \theta^2} - \frac{2}{r^2}\frac{\partial v_\theta}{\partial \theta} + \frac{\partial^2 v_r}{\partial z^2}\right) - \frac{\partial p}{\partial r} + \rho g_r$$

θ-Coordinate (tangential)

$$\rho\left(\frac{\partial v_\theta}{\partial t} + v_r\frac{\partial v_\theta}{\partial r} + \frac{v_\theta}{r}\frac{\partial v_\theta}{\partial \theta} + \frac{v_r v_\theta}{r} + v_z\frac{\partial v_\theta}{\partial z}\right)$$

$$= \mu\left(\frac{\partial}{\partial r}\left(\frac{1}{r}\frac{\partial}{\partial r}(rv_\theta)\right) + \frac{1}{r^2}\frac{\partial^2 v_\theta}{\partial \theta^2} + \frac{2}{r^2}\frac{\partial v_r}{\partial \theta} + \frac{\partial^2 v_\theta}{\partial z^2}\right) - \frac{1}{r}\frac{\partial p}{\partial \theta} + \rho g_\theta$$

z-Coordinate (axial)

$$\rho\left(\frac{\partial v_z}{\partial t} + v_r\frac{\partial v_z}{\partial r} + \frac{v_\theta}{r}\frac{\partial z}{\partial \theta} + v_z\frac{\partial v_z}{\partial z}\right)$$

$$= \mu\left(\frac{1}{r}\frac{\partial}{\partial r}\left(r\frac{\partial v_z}{\partial r}\right) + \frac{1}{r^2}\frac{\partial^2 v_z}{\partial \theta^2} + \frac{\partial^2 v_z}{\partial z^2}\right) - \frac{\partial p}{\partial z} + \rho g_z$$

Spherical coordinate system
r-Coordinate (axial)

$$\rho\left(\frac{\partial v_r}{\partial t} + v_r\frac{\partial v_r}{\partial r} + \frac{v_\theta}{r}\frac{\partial v_r}{\partial \theta} + \frac{v_\varphi}{r\sin\theta}\frac{\partial v_r}{\partial \varphi} - \frac{v_\theta^2 + v_\varphi^2}{r}\right)$$

$$= \mu\left(\frac{1}{r^2}\frac{\partial^2}{\partial r^2}(r^2 v_r) + \frac{1}{r^2\sin\theta}\frac{\partial}{\partial \theta}\left(\sin\theta\frac{\partial v_r}{\partial \theta}\right) + \frac{1}{r^2\sin^2\theta}\frac{\partial^2 v_r}{\partial \varphi^2}\right) - \frac{\partial p}{\partial r} + \rho g_r$$

θ-Coordinate (tangential)

$$\rho\left(\frac{\partial v_\theta}{\partial t} + v_r\frac{\partial v_\theta}{\partial r} + \frac{v_\theta}{r}\frac{\partial v_\theta}{\partial \theta} + \frac{v_\varphi}{r\sin\theta}\frac{\partial v_\theta}{\partial \varphi} + \frac{v_r v_\theta}{r} - \frac{v_\varphi^2\cot\theta}{r}\right)$$

$$= \mu\left(\frac{1}{r^2}\frac{\partial}{\partial r}\left(r^2\frac{\partial v_\theta}{\partial r}\right) + \frac{1}{r^2}\frac{\partial}{\partial \theta}\left(\frac{1}{\sin\theta}\frac{\partial}{\partial \theta}(v_\theta\sin\theta)\right)\right.$$

$$\left.+ \frac{1}{r^2\sin^2\theta}\frac{\partial^2 v_\theta}{\partial \varphi^2} + \frac{2}{r^2}\frac{\partial v_r}{\partial \theta} - \frac{2\cos\theta}{r^2\sin^2\theta}\frac{\partial v_\varphi}{\partial \varphi}\right) - \frac{1}{r}\frac{\partial p}{\partial \theta} + \rho g_\theta$$

φ-Coordinate (tangential)

$$\rho\left(\frac{\partial v_\varphi}{\partial t} + v_r\frac{\partial v_\varphi}{\partial r} + \frac{v_\theta}{r}\frac{\partial v_\varphi}{\partial \theta} + \frac{v_\varphi}{r\sin\theta}\frac{\partial v_\varphi}{\partial \varphi} + \frac{v_r v_\varphi}{r} - \frac{v_\varphi v_\theta\cot\theta}{r}\right)$$

$$= \mu\left(\frac{1}{r^2}\frac{\partial}{\partial r}\left(r^2\frac{\partial v_\varphi}{\partial r}\right) + \frac{1}{r^2}\frac{\partial}{\partial \theta}\left(\frac{1}{\sin\theta}\frac{\partial}{\partial \theta}(v_\varphi\sin\theta)\right)\right.$$

$$\left.+ \frac{1}{r^2\sin^2\theta}\frac{\partial^2 v_\varphi}{\partial \varphi^2} + \frac{2}{r^2\sin\theta}\frac{\partial v_r}{\partial \varphi} + \frac{2\cos\theta}{r^2\sin^2\theta}\frac{\partial v_\theta}{\partial \varphi}\right) - \frac{1}{r\sin\theta}\frac{\partial p}{\partial \varphi} + \rho g_\varphi$$

Fourier equations (thermal energy balances)

$$\rho C_p\frac{DT}{Dt} = \lambda\nabla^2 T + q$$

Rectangular coordinate system

$$\rho C_p\left(\frac{\partial T}{\partial t} + v_x\frac{\partial T}{\partial x} + v_y\frac{\partial T}{\partial y} + v_z\frac{\partial T}{\partial z}\right) = \lambda\left(\frac{\partial^2 T}{\partial x^2} + \frac{\partial^2 T}{\partial y^2} + \frac{\partial^2 T}{\partial z^2}\right) + q$$

Cylindrical coordinate system

$$\rho C_p\left(\frac{\partial T}{\partial t} + v_r\frac{\partial T}{\partial r} + \frac{v_\theta}{r}\frac{\partial T}{\partial \theta} + v_z\frac{\partial T}{\partial z}\right) = \lambda\left(\frac{1}{r}\frac{\partial}{\partial r}\left(r\frac{\partial T}{\partial r}\right) + \frac{1}{r^2}\frac{\partial^2 T}{\partial \theta^2} + \frac{\partial^2 T}{\partial z^2}\right) + q$$

Spherical coordinate system

$$\rho C_p\left(\frac{\partial T}{\partial t} + v_r\frac{\partial T}{\partial r} + \frac{v_\theta}{r}\frac{\partial T}{\partial \theta} + \frac{v_\varphi}{r\sin\theta}\frac{\partial T}{\partial \varphi}\right)$$

$$= \lambda\left(\frac{1}{r^2}\frac{\partial}{\partial r}\left(r^2\frac{\partial T}{\partial r}\right) + \frac{1}{r^2\sin\theta}\frac{\partial}{\partial \theta}\left(\sin\theta\frac{\partial T}{\partial \theta}\right) + \frac{1}{r^2\sin^2\theta}\frac{\partial^2 T}{\partial \varphi^2}\right) + q$$

Fick's equations (component mass balances)

$$\frac{Dc_A}{Dt} = \mathcal{D}_{AB}\ \nabla^2 c_A + \mathcal{R}_A$$

Rectangular coordinate system

$$\left(\frac{\partial c_A}{\partial t} + v_x \frac{\partial c_A}{\partial x} + v_y \frac{\partial c_A}{\partial y} + v_z \frac{\partial c_A}{\partial z}\right) = \mathcal{D}_{AB}\left(\frac{\partial^2 c_A}{\partial x^2} + \frac{\partial^2 c_A}{\partial y^2} + \frac{\partial^2 c_A}{\partial z^2}\right) + \mathcal{R}_A$$

Cylindrical coordinate system

$$\left(\frac{\partial c_A}{\partial t} + v_r \frac{\partial c_A}{\partial r} + \frac{v_\theta}{r}\frac{\partial c_A}{\partial \theta} + v_z \frac{\partial c_A}{\partial z}\right) = \mathcal{D}_{AB}\left(\frac{1}{r}\frac{\partial}{\partial r}\left(r\frac{\partial c_A}{\partial r}\right) + \frac{1}{r^2}\frac{\partial^2 c_A}{\partial \theta^2} + \frac{\partial^2 c_A}{\partial z^2}\right) + \mathcal{R}_A$$

Spherical coordinate system

$$\left(\frac{\partial c_A}{\partial t} + v_r \frac{\partial c_A}{\partial r} + \frac{v_\theta}{r}\frac{\partial c_A}{\partial \theta} + \frac{v_\varphi}{r\sin\theta}\frac{\partial c_A}{\partial \varphi}\right)$$

$$= \mathcal{D}_{AB}\left(\frac{1}{r^2}\frac{\partial}{\partial r}\left(r^2\frac{\partial c_A}{\partial r}\right) + \frac{1}{r^2\sin\theta}\frac{\partial}{\partial \theta}\left(\sin\theta\,\frac{\partial c_A}{\partial \theta}\right) + \frac{1}{r^2\sin^2\theta}\frac{\partial^2 c_A}{\partial \varphi^2}\right) + \mathcal{R}_A$$

Continuity equation and momentum balance for special cases

Continuity equation for compressible flow

$$\frac{D}{Dt}(\rho v) = 0$$

Rectangular coordinate system

$$\frac{\partial \rho}{\partial t} + \frac{\partial \rho v_x}{\partial x} + \frac{\partial \rho v_y}{\partial y} + \frac{\partial \rho v_z}{\partial z} = 0$$

Cylindrical coordinate system

$$\frac{\partial \rho}{\partial t} + \frac{1}{r}\frac{\partial (r \rho v_r)}{\partial r} + \frac{1}{r}\frac{\partial \rho v_\theta}{\partial \theta} + \frac{\partial \rho v_z}{\partial z} = 0$$

Spherical coordinate system

$$\frac{\partial \rho}{\partial t} + \frac{1}{r^2}\frac{\partial}{\partial r}\left(r^2 \rho v_r\right) + \frac{1}{r\,\sin\theta}\frac{\partial}{\partial \theta}\left(\rho v_\theta\,\sin\theta\right) + \frac{1}{r\,\sin\theta}\frac{\partial \rho v_\phi}{\partial \varphi} = 0$$

Momentum balance for incompressible non-Newtonian fluids

$$\rho\frac{Dv}{Dt} = -\nabla\cdot\tau - \nabla p + \rho g$$

Rectangular coordinate system
x-Coordinate

$$\rho\left(\frac{\partial v_x}{\partial t} + v_x\frac{\partial v_x}{\partial x} + v_y\frac{\partial v_x}{\partial y} + v_z\frac{\partial v_x}{\partial z}\right) = -\left(\frac{\partial \tau_{xx}}{\partial x} + \frac{\partial \tau_{yx}}{\partial y} + \frac{\partial \tau_{zx}}{\partial z}\right) - \frac{\partial p}{\partial x} + \rho g_x$$

y-Coordinate

$$\rho\left(\frac{\partial v_y}{\partial t} + v_x\frac{\partial v_y}{\partial x} + v_y\frac{\partial v_y}{\partial y} + v_z\frac{\partial v_y}{\partial z}\right) = -\left(\frac{\partial \tau_{xy}}{\partial x} + \frac{\partial \tau_{yy}}{\partial y} + \frac{\partial \tau_{zy}}{\partial z}\right) - \frac{\partial p}{\partial y} + \rho g_y$$

z-Coordinate

$$\rho\left(\frac{\partial v_z}{\partial t} + v_x\frac{\partial v_z}{\partial x} + v_y\frac{\partial v_z}{\partial y} + v_z\frac{\partial v_z}{\partial z}\right) = -\left(\frac{\partial \tau_{xz}}{\partial x} + \frac{\partial \tau_{yz}}{\partial y} + \frac{\partial \tau_{zz}}{\partial z}\right) - \frac{\partial p}{\partial z} + \rho g_z$$

Cylindrical coordinate system
r-Coordinate (radial)

$$\rho\left(\frac{\partial v_r}{\partial t} + v_r\frac{\partial v_r}{\partial r} + \frac{v_\theta}{r}\frac{\partial v_r}{\partial \theta} - \frac{v_\theta^2}{r} + v_z\frac{\partial v_r}{\partial z}\right)$$

$$= -\left(\frac{1}{r}\frac{\partial}{\partial r}(r\tau_{rr}) + \frac{1}{r}\frac{\partial \tau_{r\theta}}{\partial \theta} - \frac{\tau_{\theta\theta}}{r} + \frac{\partial \tau_{rz}}{\partial z}\right) - \frac{\partial p}{\partial r} + \rho g_r$$

θ-Coordinate (tangential)

$$\rho\left(\frac{\partial v_\theta}{\partial t} + v_r\frac{\partial v_\theta}{\partial r} + \frac{v_\theta}{r}\frac{\partial v_\theta}{\partial \theta} + \frac{v_r v_\theta}{r} + v_z\frac{\partial v_\theta}{\partial z}\right)$$

$$= -\left(\frac{1}{r^2}\frac{\partial}{\partial r}(r^2\tau_{r\theta}) + \frac{1}{r}\frac{\partial \tau_{\theta\theta}}{\partial \theta} + \frac{\partial \tau_{\theta z}}{\partial z} + \frac{\tau_{\theta r} - \tau_{r\theta}}{\tau}\right) - \frac{1}{r}\frac{\partial p}{\partial \theta} + \rho g_\theta$$

z-Coordinate (axial)

$$\rho\left(\frac{\partial v_z}{\partial t} + v_r\frac{\partial v_z}{\partial r} + \frac{v_\theta}{r}\frac{\partial v_z}{\partial \theta} + v_z\frac{\partial v_z}{\partial z}\right)$$

$$= -\left(\frac{1}{r}\frac{\partial}{\partial r}(r\tau_{rz}) + \frac{1}{r}\frac{\partial \tau_{\theta z}}{\partial \theta} + \frac{\partial \tau_{zz}}{\partial z}\right) - \frac{\partial p}{\partial z} + \rho g_z$$

Spherical coordinate system
r-Coordinate (axial)

$$\rho\left(\frac{\partial v_r}{\partial t} + v_r\frac{\partial v_r}{\partial r} + \frac{v_\theta}{r}\frac{\partial v_r}{\partial \theta} + \frac{v_\varphi}{r\sin\theta}\frac{\partial v_r}{\partial \varphi} - \frac{v_\theta^2 + v_\varphi^2}{r}\right)$$

$$= -\left(\frac{1}{r^2}\frac{\partial}{\partial r}(r^2\tau_{rr}) + \frac{1}{r\sin\theta}\frac{\partial}{\partial \theta}(\tau_{\theta r}\sin\theta) + \frac{1}{r\sin\theta}\frac{\partial \tau_{r\varphi}}{\partial \varphi} - \frac{\tau_{\theta\theta} + \tau_{\varphi\varphi}}{r}\right) - \frac{\partial p}{\partial r} + \rho g_r$$

θ-Coordinate (tangential)

$$\rho\left(\frac{\partial v_\theta}{\partial t} + v_r\frac{\partial v_\theta}{\partial r} + \frac{v_\theta}{r}\frac{\partial v_\theta}{\partial \theta} + \frac{v_\varphi}{r\sin\theta}\frac{\partial v_\theta}{\partial \varphi} + \frac{v_r v_\theta}{r} + \frac{v_\varphi^2\cot\theta}{r}\right)$$

$$= -\left(\frac{1}{r^2}\frac{\partial}{\partial r}(r^2\tau_{r\theta}) + \frac{1}{r\sin\theta}\frac{\partial}{\partial \theta}(\tau_{\theta\theta}\sin\theta) + \frac{1}{r\sin\theta}\frac{\partial \tau_{\theta\varphi}}{\partial \varphi} + \frac{\tau_{r\theta}}{r} - \frac{\cot\theta}{r}\tau_{\varphi\varphi}\right) - \frac{1}{r}\frac{\partial p}{\partial \theta} + \rho g_\theta$$

φ-Coordinate (tangential)

$$\rho\left(\frac{\partial v_\varphi}{\partial t} + v_r\frac{\partial v_\varphi}{\partial r} + \frac{v_\theta}{r}\frac{\partial v_\varphi}{\partial \theta} + \frac{v_\varphi}{r\sin\theta}\frac{\partial v_\varphi}{\partial \varphi} + \frac{v_r v_\varphi}{r} + \frac{v_\varphi v_\theta \cot\theta}{r}\right)$$

$$= -\left(\frac{1}{r^2}\frac{\partial}{\partial r}\left(r^2\tau_{r\phi}\right) + \frac{1}{r}\frac{\partial\tau_{\theta\varphi}}{\partial\theta} + \frac{1}{r\sin\theta}\frac{\partial\tau_{\varphi\varphi}}{\partial\varphi} + \frac{\tau_{r\varphi}}{r} + \frac{2\cot\theta}{r}\,\tau_{\theta\varphi}\right) - \frac{1}{r\sin\theta}\frac{\partial p}{\partial\varphi} + \rho g_\varphi$$

List of symbols and abbreviations

c_A	Concentration of component A
C_p	Specific heat
$\mathcal{D}_{AB}$	Diffusion coefficient of component A in B
D/Dt	Material derivative
g_i	Gravity in the direction i
p	Pressure
q	Heat production per unit volume
$\mathcal{R}_A$	Production of component A
T	Temperature
t	Time
$\boldsymbol{v}$	Velocity vector
v_i	Velocity in the i direction
$x, y, z, \varphi, \theta, r$	Coordinates
$\partial/\partial t$	Partial derivative
λ	Heat conductivity
μ	Viscosity
ρ	Density
$\boldsymbol{\tau}$	Shear stress tensor
τ_{ij}	Shear stress in directions i, j
∇	Scalar gradient or covariant derivative of a vector

Macrobalances

Unsteady-state balances

Macroscopic mass balance

$$\frac{dm_{\text{tot}}}{dt} = -\Delta w = -\Delta(\rho\langle v\rangle S)$$

Macroscopic momentum balance

$$\frac{dP_{\text{tot}}}{dt} = -\Delta\left(\frac{\langle v^2\rangle}{\langle v\rangle}w + pS\right) - F + m_{\text{tot}}g$$

Macroscopic energy balance

$$\frac{dE_{\text{tot}}}{dt} = -\Delta\left\{\left(\hat{U} + p\hat{V} + \frac{1}{2}\frac{\langle v^3\rangle}{\langle v\rangle} + gH\right)w\right\} + Q - W$$

Macroscopic mechanical energy balance

$$\frac{dE_{\text{tot}}}{dt} = -\left(\int_{p_1}^{p_2}\frac{1}{\rho}dp + \Delta\left(\frac{1}{2}\frac{\langle v^3\rangle}{\langle v\rangle}\right) + \Delta(gH)\right)w - E_v - W$$

Steady-state balances

Macroscopic mass balance

$$\Delta w = \Delta(\rho\langle v\rangle S) = 0$$

Macroscopic momentum balance

$$F = -\Delta\left(\frac{\langle v^2\rangle}{\langle v\rangle}w + pS\right) + m_{tot}g$$

Macroscopic energy balance

$$\Delta\left\{\left(\hat{U} + p\hat{V} + \frac{1}{2}\frac{\langle v^3\rangle}{\langle v\rangle} + gH\right)w\right\} = Q - W$$

Macroscopic mechanical energy balance (Bernoulli's equation)

$$\left(\int_{p_1}^{p_2} \frac{1}{\rho} dp + \Delta \left(\frac{1}{2} \frac{\langle v^3 \rangle}{\langle v \rangle} \right) + \Delta(gH) \right) w + E_v + W = 0$$

Symbol list

Symbol		Unit
E_{tot}	Total energy	J
E_v	Friction losses	W
F	Force	N
g	Gravitational constant	m/s^2
m_{tot}	Total mass	kg
P_{total}	Total momentum	kg m/s
p	Pressure	Pa
Q	Thermal energy added	W
S	Surface	m^2
t	Time	s
$\hat{U}$	Internal energy per unit mass	m^2/s^2
v	Velocity	m/s
$\hat{V}$	Volume per unit mass	m^3/kg
w	Mass flow	kg/s
W	Work done on surroundings	W
ρ	Density	kg/m^3

Elementary differential equations for flow, heat transfer, and mass transfer

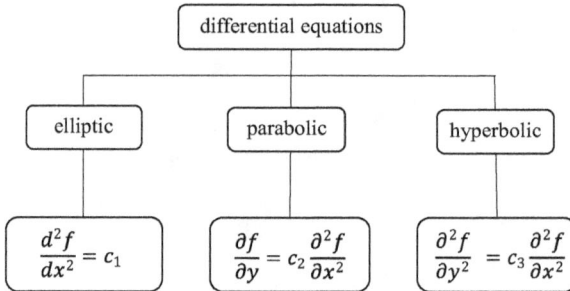

Fig. 4.1: Different types of differential equations.

Fluid flow

Gravity flow along a flat plate

For example, film flow over a vertical plate

$$\mu \frac{\partial^2 v_z}{\partial x^2} + \rho g_z = 0$$

Boundary conditions: $\quad x = \delta \rightarrow \dfrac{\partial v_z}{\partial x} = 0$

$$x = 0 \rightarrow v_z = 0$$

$$v_z = \frac{\rho g_z}{\mu} \left(x\delta - \frac{x^2}{2} \right)$$

$$\Phi_v = \frac{\rho g \delta^3}{3\mu} W$$

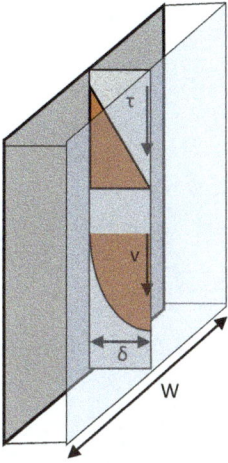

Fig. 4.2: Film flow along a wall.

Rectangular coordinate system with parallel pressure gradient

For example, pressure flow between flat plates

$$\frac{\partial p}{\partial z} + \frac{\partial \tau_{xz}}{\partial x} = \frac{\partial p}{\partial z} + \mu \left(\frac{\partial^2 v_z}{\partial x^2} \right) = 0$$

Boundary conditions: $x = 0 \rightarrow \dfrac{\partial v_z}{\partial x} = 0$

$$x = \pm d \rightarrow v_z = 0$$

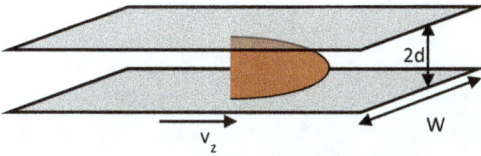

Fig. 4.3: Pressure flow between flat plates.

$$v_z = \frac{1}{2\mu} \left(-\frac{dp}{dz} \right) (d^2 - x^2)$$

$$\Phi_v = \frac{2W}{3\mu} \left(-\frac{dp}{dz} \right) d^3$$

Cylindrical coordinate system with axial pressure gradient

For example, laminar flow in a round tube

$$\frac{\partial p}{\partial z} + \frac{1}{r}\frac{\partial}{\partial r}(r\tau_{rz}) = \frac{\partial p}{\partial z} - \mu\frac{1}{r}\frac{\partial}{\partial r}\left(r\frac{\partial v_z}{\partial r}\right) = 0$$

Boundary conditions: $\quad r = 0 \rightarrow \dfrac{\partial v_z}{\partial r} = 0$

$$r = R \rightarrow v_z = 0$$

$$v_z = \frac{1}{4\mu}\left(-\frac{dp}{dz}\right)(R^2 - r^2)$$

$$\Phi_v = \frac{\pi R^4}{8\,\mu}\left(-\frac{dp}{dz}\right)$$

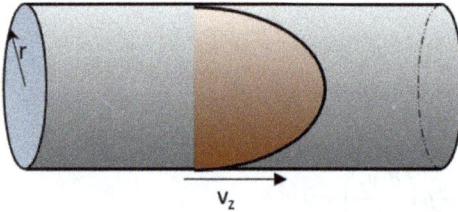

Fig. 4.4: Laminar pipe flow.

Rectangular coordinate system with axial pressure gradient

For example, laminar flow in a tube with square cross section and sides of a

$$\frac{\partial p}{\partial z} + \frac{\partial \tau_{xz}}{\partial x} + \frac{\partial \tau_{yz}}{\partial y} = \frac{\partial p}{\partial z} - \mu\left(\frac{\partial^2 v_z}{\partial x^2}\right) - \mu\left(\frac{\partial^2 v_z}{\partial y^2}\right) = 0$$

Boundary conditions: $\quad x = 0 \text{ and } x = a \rightarrow v_z = 0$

$$y = 0 \text{ and } y = a \rightarrow v_z = 0$$

$$\Phi_v = \frac{a^4}{28.6\,\mu}\left(-\frac{dp}{dz}\right)$$

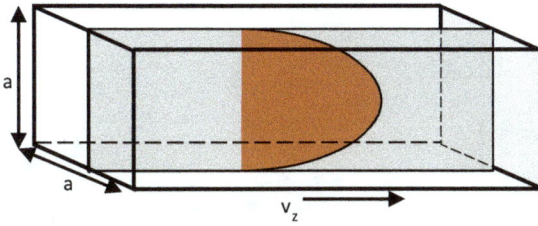

Fig. 4.5: Flow in a square pipe.

Cylindrical coordinate system with axial pressure gradient and power function fluid

For example, flow of a power law fluid in a tube

$$\frac{\partial p}{\partial z} + \frac{1}{r}\frac{\partial}{\partial r}\left(r\tau_{rz}\right) = \frac{\partial p}{\partial z} + k\frac{2}{r}\frac{\partial}{\partial r}\left(\left[\frac{dv_z}{dr}\right]^n\right) = 0$$

Boundary conditions: $r = 0 \rightarrow \dfrac{\partial v_z}{\partial r} = 0$

$r = R \rightarrow v_z = 0$

$$v_z = \frac{n}{n+1}\left\{\frac{1}{2k}\left(-\frac{dp}{dz}\right)\right\}^{\frac{1}{n}}\left(R^{\frac{n+1}{n}} - r^{\frac{n+1}{n}}\right)$$

$$\Phi_v = \frac{\pi n}{3n+1}\left\{\frac{1}{2k}\left(-\frac{dp}{dz}\right)\right\}^{\frac{1}{n}} R^{\frac{3n+1}{n}}$$

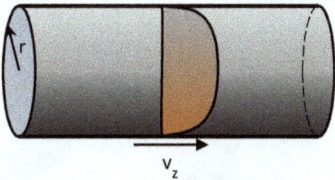

Fig. 4.6: Flow of a power law fluid in a pipe.

Cylindrical coordinate system with tangential flow

For example, flow in a concentric bearing

$$0 = \frac{\delta}{\delta r}\left(\frac{1}{r}\frac{\partial(rv_\theta)}{\partial r}\right)$$

Boundary conditions: $\qquad r = R_i \rightarrow v_\theta = 0$

$$r = R_o \rightarrow v_\theta = \omega R_0$$

$$v_\theta = \omega R_0 \dfrac{\frac{R_i}{r} - \frac{r}{R_i}}{\frac{R_i}{R_0} - \frac{R_0}{R_i}}$$

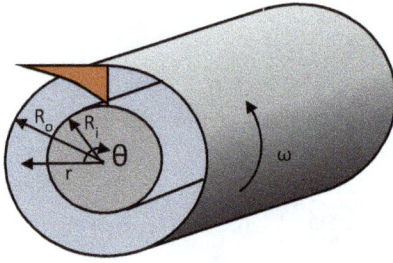

Fig. 4.7: Tangential flow between concentric cyliders.

Cylindrical coordinate system with axial flow

$$\frac{1}{r}\frac{d}{dr}(r\tau_{rz}) = -\frac{\mu}{r}\frac{d}{dr}\left(r\frac{dv_z}{dr}\right) = -\frac{dp}{dz}$$

Boundary conditions: $\qquad r = R_i \rightarrow v_z = 0$

$$r = R_o \rightarrow v_z = 0$$

$$v_z = -\frac{R_0^2}{4\mu}\frac{dp}{dz}\left\{1 - \left(\frac{r}{R_0}\right)^2 + \left(\frac{1 - (R_i/R_0)^2}{\ln(R_0/R_i)}\right)\ln\left(\frac{r}{R_0}\right)\right\}$$

$$\Phi_v = -\frac{\pi R_0^4}{8\mu}\frac{dp}{dz}\left\{\left(\left(1 - (R_i/R_0)^4 - \frac{\left(1 - (R_i/R_0)^2\right)^2}{\ln(R_0/R_i)}\right)\right)\right\}$$

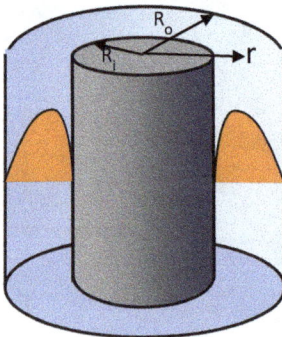

Fig. 4.8: Axial flow between concentric cylinders.

Heat transfer

Sphere with uniform internal heat production, Bi $\gg$1 ($h_{int} \ll h_{ext}$)

For example, catalyst particle, bio-pellet

$$\lambda\left(\frac{1}{r^2}\frac{\partial}{\partial r}\left(r^2\frac{\partial T}{\partial r}\right)\right) + Q = 0$$

Boundary conditions: $r = 0 \rightarrow \dfrac{dT}{dr} = 0$

$$r = R \rightarrow T = T_w$$

$$T - T_w = \frac{Q}{6\lambda}\left(R^2 - r^2\right)$$

$$\Phi_h = Q \times V = Q \times \frac{4}{3}\pi R^3$$

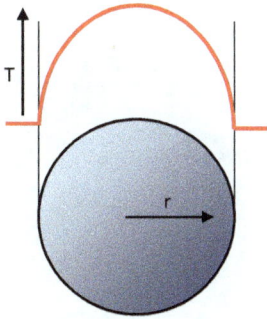

Fig. 4.9: Temperature profile in a sphere.

Heating of infinite surroundings by a sphere at constant temperature (Bi $\ll$ 1) ($h_{int} \gg h_{ext}$)

For example, stagnant surroundings of a metal sphere

$$\frac{\partial}{\partial r}\left(r^2\frac{\partial T}{\partial r}\right) = 0$$

Boundary conditions: $r = R \rightarrow T = T_w$

$$r \rightarrow \infty \rightarrow T = T_\infty$$

$$\frac{T - T_\infty}{T_w - T_\infty} = \frac{R}{r}$$

$$\Phi_h = 4\pi R\lambda(T_w - T_\infty)$$

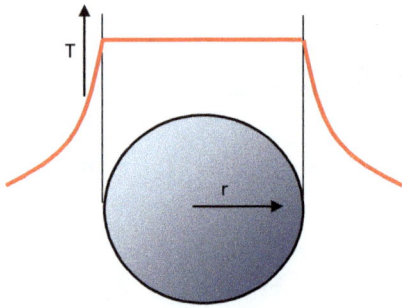

Fig. 4.10: Temperature profile around a sphere.

Cylinder with uniform internal heating (Bi $\gg$ 1) ($h_{int} \ll h_{ext}$)
For example, fuel element in a nuclear reactor

$$0 = \lambda \frac{1}{r}\frac{\partial}{\partial r}\left(r\frac{\partial T}{\partial r}\right) + Q$$

Boundary conditions: $\quad r = 0 \rightarrow \dfrac{dT}{dr} = 0$

$\qquad\qquad\qquad\quad r = R \rightarrow T = T_w.$

$$T - T_w = \frac{Q}{4\lambda}\left(R^2 - r^2\right)$$

$$\Phi_h = Q \times V = Q \times \pi R^2\, L$$

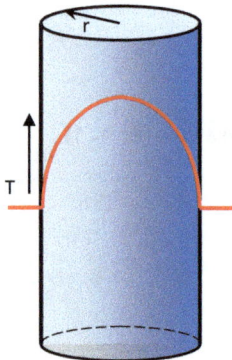

Fig. 4.11: Temperature profile in a cylinder with heat generation.

Annulus with temperature difference

For example, isolation of a round tube

$$\frac{\partial}{\partial r}\left(r\frac{\partial T}{\partial r}\right) = 0$$

Boundary conditions: $r = R_1 \rightarrow T = T_1$

$r = R_2 \rightarrow T = T_2$

$$\frac{T - T_1}{T_2 - T_1} = \frac{\ln(r/R_1)}{\ln(R_2/R_1)} \text{ or } \frac{T - T_2}{T_1 - T_2} = \frac{\ln(r/R_2)}{\ln(R_1/R_2)}$$

$$\Phi_h = \frac{-2\pi\lambda L}{\ln(R_1/R_2)}(T_1 - T_2) = \frac{-2\pi\lambda L}{\ln(R_2/R_1)}(T_2 - T_1)$$

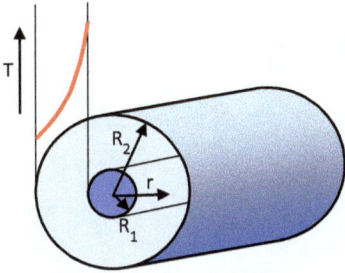

Fig. 4.12: Temperature profile in annular isolation.

Heat penetration in a semi-infinite body ($t \ll d^2/a$)

For example, short time heat penetration

$$\frac{\partial T}{\partial t} = a\frac{\partial^2 T}{\partial x^2}$$

Boundary conditions: $t \leq 0 \rightarrow T = T_0$

$t > 0, x = 0 \rightarrow T = T_1$

$t > 0, x \rightarrow \infty \rightarrow T = T_0$

$$\frac{T_1 - T}{T_1 - T_0} = \text{erf}\left(\frac{x}{2\sqrt{at}}\right) \quad \left(\text{Fo} = \frac{at}{d^2} \ll 1\right)$$

$$\Phi_h'' = (T_1 - T_0)\sqrt{\frac{\lambda\rho C_p}{\pi t}}$$

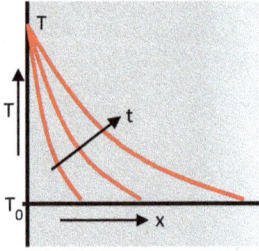

Fig. 4.13: Short time temperature changes.

Two-sided heating of a flat plate (long time approximation ($t > d^2/a$))
For example, heating of a plate after long time

$$\frac{\partial T}{\partial t} = a\frac{\partial^2 T}{\partial x^2}$$

Boundary conditions: $t \leq 0 \rightarrow T = T_0$

$$t > 0, x = 0 \rightarrow \frac{\partial T}{\partial x} = 0$$

$$t > 0, x = \pm\, d \rightarrow T = T_1$$

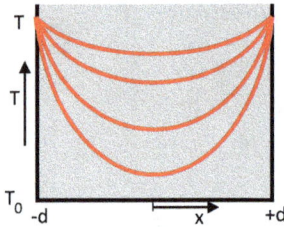

Fig. 4.14: Long time temperature changes.

$$\frac{T_1 - T}{T_1 - T_0} = 2\sum_{n=0}^{\infty} \frac{(-1)^n}{(n+\frac{1}{2})\pi} \exp\left(-\left(n+\frac{1}{2}\right)^2 \frac{\pi^2 at}{d^2}\right) \cos\left(\left(n+\frac{1}{2}\right)\frac{\pi x}{d}\right) \quad \left(Fo = \frac{at}{d^2} > 1\right)$$

Approximation *if* Fo $\gg$ 1:

$$\frac{T_1 - T}{T_1 - T_0} = \frac{4}{\pi}\exp\left(-\frac{\pi^2 at}{4d^2}\right)\cos\left(\frac{\pi x}{2d}\right)$$

$$\Phi_h^{''} = \frac{4}{d}\exp\left(-\frac{\pi^2 at}{4d^2}\right)(T_1 - T_0)$$

Heat transfer at a flat plate

For example, rectangular cooling fin

$$\frac{d^2 T}{dz^2} = \frac{h}{\lambda B}(T - T_0)$$

Boundary conditions: $z = 0 \rightarrow T = T_w$

$$z = L \rightarrow \frac{dT}{dz} = 0$$

$$\frac{T - T_0}{T_w - T_0} = \frac{\cosh\left\{\sqrt{\frac{h}{\lambda B}}(L - z)\right\}}{\cosh\sqrt{\frac{hL^2}{\lambda B}}}$$

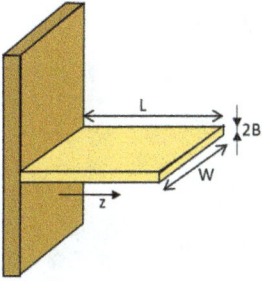

Fig. 4.15: Horizontal cooling fin.

Mass transfer

Sphere with uniform internal production of A $\left(\dfrac{\mathcal{D}_A}{d} \ll k_{ext}\right)$

For example, bio-pellet with zero-th order reaction

$$\mathcal{D}_A\left(\frac{1}{r^2}\frac{\partial}{\partial r}\left(r^2\frac{\partial C_A}{\partial r}\right)\right) + R_A = 0$$

Boundary conditions: $r = 0 \rightarrow \dfrac{dC_A}{dr} = 0$

$$r = R \rightarrow C_A = C_{Aw}$$

$$C_A - C_{Aw} = \frac{R_A}{6\mathcal{D}_A}(R^2 - r^2)$$

$$\Phi_{C_A} = R_A \times V = R_A \times \frac{4}{3}\pi R^3$$

$$C_{Aw} = mC_{A\infty}$$

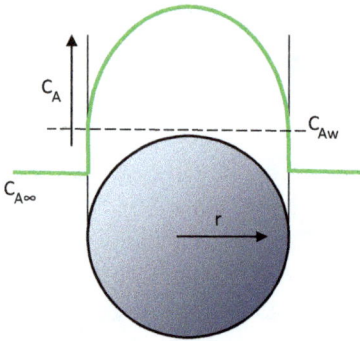

Fig. 4.16: Concentration profile in a sphere.

Diffusion through infinite surroundings of a sphere $\left(\dfrac{\mathcal{D}_A}{d} \ll k_{\text{int}} \right)$

For example, dissolution of a sphere (constant radius) in a stationary gel

$$\frac{\partial}{\partial r}\left(r^2 \frac{\partial C_A}{\partial r} \right) = 0$$

Boundary conditions: $r = R \rightarrow C_A = C_{Aw}$

$r \rightarrow \infty \rightarrow C_A = C_{A\infty}$

$$\frac{C_A - C_{A\infty}}{C_{Aw} - C_{A\infty}} = \frac{R}{r}$$

$$\Phi_{C_A} = 4\pi R \mathcal{D}_A (C_{Aw} - C_{A\infty})$$

$$C_{Aw} = m C_{A,\text{int}}$$

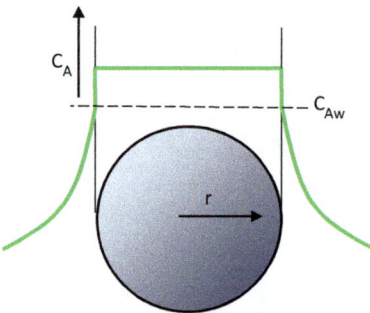

Fig. 4.17: Concentration profile around a sphere.

Cylinder with uniform internal production of A $\left(\dfrac{\mathcal{D}_A}{d} \ll k_{\text{ext}} \right)$

For example, cylindrical bioreactor with zero-th order reaction

$$0 = \mathcal{D}_A \frac{1}{r} \frac{\partial}{\partial r} \left(r \frac{\partial C_A}{\partial r} \right) + R_A$$

Boundary conditions: $\quad r = 0 \rightarrow \dfrac{dC_A}{dr} = 0;$

$$r = R \rightarrow C_A = C_{Aw}.$$

$$C_A - C_{Aw} = \frac{R_A}{4\mathcal{D}_A} \left(R^2 - r^2 \right)$$

$$\Phi_A = R_A \times V = R_A \times \pi R^2 \, L$$

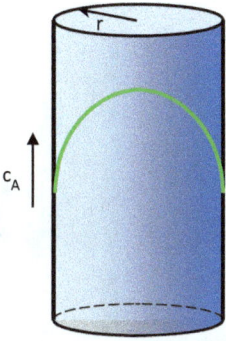

Fig. 4.18: Concentration profile in a cylinder.

Diffusion between annular surfaces

For example, chemical vapor deposition in a cylindrical cavity

$$\frac{\partial}{\partial r} \left(r \frac{\partial C_A}{\partial r} \right) = 0$$

Boundary conditions: $\quad r = R_1 \rightarrow C_A = C_{A1}$

$$r = R_2 \rightarrow C_A = C_{A2}$$

$$\frac{C_A - C_{A1}}{C_{A2} - C_{A1}} = \frac{\ln(r/R_1)}{\ln(R_2/R_1)} \quad \text{or} \quad \frac{C_A - C_{A2}}{C_{A1} - C_{A2}} = \frac{\ln(r/R_2)}{\ln(R_1/R_2)}$$

$$\Phi_A = \frac{-2\pi \mathcal{D}L}{\ln(R_1/R_2)} (C_{A1} - C_{A2}) = \frac{-2\pi \mathcal{D}L}{\ln(R_2/R_1)} (C_{A2} - C_{A1})$$

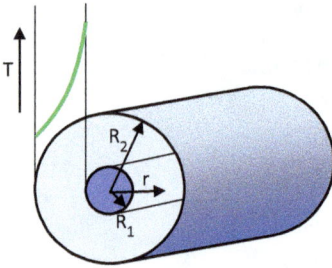

Fig. 4.19: Concentration profile in an annulus.

Instationary diffusion, short times, or infinite medium ($t \ll d^2/\mathcal{D}_A$)
For example, penetration of a contaminant into the soil

$$\frac{\partial C_A}{\partial t} = \mathcal{D}_A \frac{\partial^2 C_A}{\partial x^2}$$

Boundary conditions: $t \leq 0 \rightarrow C_A = C_{A0}$

$t > 0, x = 0 \rightarrow C_A = C_{Aw}$

$t > 0, x \rightarrow \infty \rightarrow C_A = C_{A0}$

$$\frac{C_{Aw} - C_A}{C_{Aw} - C_{A0}} = \mathrm{erf}\left(\frac{x}{2\sqrt{\mathcal{D}_A t}}\right) \left(\mathrm{Fo} = \frac{\mathcal{D}t}{d^2} \ll 1\right)$$

$$\Phi_A'' = (C_{Aw} - C_{A0})\sqrt{\frac{\mathcal{D}_A}{\pi t}}$$

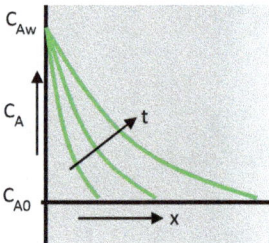

Fig. 4.20: Short time concentration changes.

Instationary diffusion in a finite medium after longer times $(t > d^2/\mathcal{D}_A)$
For example, concentration equalization flat plate

$$\frac{\partial C_A}{\partial t} = \mathcal{D}_A \frac{\partial^2 C_A}{\partial x^2}$$

Boundary conditions: $\quad t \leq 0 \rightarrow C_A = C_{A0}$

$$t > 0, x = 0 \rightarrow \frac{\partial C_A}{\partial x} = 0$$

$$t > 0, x = \pm\, d \rightarrow C_A = C_{Aw}$$

$$\frac{C_{Aw} - C_A}{C_{Aw} - C_{A0}} = 2 \sum_{n=0}^{\infty} \frac{(-1)^n}{(n + \frac{1}{2})\pi} \exp\left(-\left(n + \frac{1}{2}\right)^2 \frac{\pi^2 \mathcal{D}_A t}{d^2}\right) \cos\left(\left(n + \frac{1}{2}\right)\frac{\pi x}{d}\right)$$

$$\left(\mathrm{Fo} = \frac{\mathcal{D}_A t}{d^2} > 1\right)$$

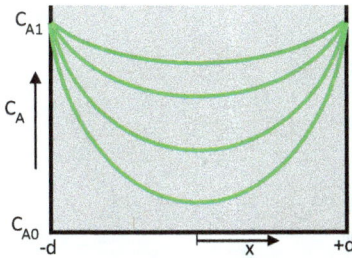

Fig. 4.21: Long time concentration changes.

Approximation if $(\mathrm{Fo} \gg 1)$:

$$\frac{C_{Aw} - C_A}{C_{Aw} - C_{A0}} = \frac{4}{\pi} \exp\left(-\frac{\pi^2 \mathcal{D}_A t}{4d^2}\right) \cos\left(\frac{\pi x}{2d}\right)$$

$$\Phi_A'' = \frac{4}{d} \exp\left(-\frac{\pi^2 \mathcal{D}_A t}{4d^2}\right)(C_{Aw} - C_{A0})$$

List of symbols and abbreviations

a	Side of a square	m
a	Heat diffusivity	m²/s
Bi	Biot number	–
C_A	Concentration of component A	kmol/m³ or kg/m³
C_{Aw}	Concentration of A at the wall	kmol/m³ or kg/m³
C_{A0}	Concentration of A at $t = 0$	kmol/m³ or kg/m³
c_i	Constant (i = 1, 2, 3, . . .)	–
const	Integration constant	
d	Thickness	m
$\mathcal{D}_A$	Diffusion coefficient of component A	m²/s
f, g	Arbitrary functions	
Fo	Fourier number	–

g_i	Gravitation constant in direction i (x, y, z)	m/s^2
h	Heat transfer coefficient	$W/m^2\ K$
k	Power law consistency	$Pa\ s^n$
k	Mass transfer coefficient	m^2/s
L	Length	m
n	Power law index	–
n	Order of the Bernoulli equation	–
p	Pressure	Pa
Q	Specific heat production	W/m^3
R	Radius	m
r	Radial coordinate	m
R_A	Production of compound A by reaction	$kmol/m^3\ s$ or $kg/m^3\ s$
R_i	Inner radius	m
R_o	Outer radius	m
T	Temperature	°C or K
t	Time	s
T_∞	Temperature at infinity	°C or K
T_w	Wall temperature	°C or K
V	Volume	m^3
v_i	Velocity $(i = x, y, z, \theta, r)$	m/s
W	Width	m
x, y, z	Coordinates	m
δ	Film thickness	m
θ	Tangential coordinate	rad
λ	Heat conductivity	$W/m\ K$
μ	Viscosity	Pa s
ρ	Density	kg/m^3
$\tau_{i,j}$	Shear stress $(i, j = x, y, z, \theta, r)$	Pa
ω	Rotation rate	1/s
Φ_A	Molar or mass flow of comp. A	kmol/s or kg/s
Φ_v''	Molar or mass flow rate of comp. A	$kmol/m^2\ s$ or $kg/m^2\ s$
Φ_v	Volumetric flow	m^3/s
Φ_v''	Volumetric flow rate	m/s
Φ_h	Heat flow	W
Φ_h''	Heat flow rate	W/m^2

Frequently occurring flow fields in tensor notation

Notation: $\dot{\gamma} = \dfrac{\partial v_x}{\partial y}$; $\dot{\varepsilon} = \dfrac{\partial v_x}{\partial x}$; $\underline{\underline{\gamma}}$ deformation tensor

Simple shear

$$\underline{\underline{\dot{\gamma}}} = \begin{pmatrix} 0 & 1 & 0 \\ 1 & 0 & 0 \\ 0 & 0 & 0 \end{pmatrix} \quad \dot{\gamma}$$

Simple elongation

Elongation in the x-direction, no forces in the y- and z-directions

$$\underline{\underline{\dot{\gamma}}} = \begin{pmatrix} 1 & 0 & 0 \\ 0 & -\frac{1}{2} & 0 \\ 0 & 0 & -\frac{1}{2} \end{pmatrix} \quad \dot{\varepsilon}$$

Uniaxial elongation

Elongation in the x-direction, no forces in the y-direction, and no velocities in the z-direction

$$\underline{\underline{\dot{\gamma}}} = \begin{pmatrix} 1 & 0 & 0 \\ 0 & -1 & 0 \\ 0 & 0 & 0 \end{pmatrix} \quad \dot{\varepsilon}$$

Biaxial elongation

Equal elongation in the x- and y-directions, no forces in the z-direction

$$\underline{\underline{\dot{\gamma}}} = \begin{pmatrix} 1 & 0 & 0 \\ 0 & 1 & 0 \\ 0 & 0 & -2 \end{pmatrix} \quad \dot{\varepsilon}$$

Rheological models

Notation

One-dimensional	τ: shear stress	A, B, C: constants
	η: viscosity	$\dot{\gamma} = \dfrac{dv_x}{dy}$
	τ_0: yield stress	
	k: consistency	$\eta_\infty = \lim\limits_{\dot{\gamma} \to 0} \eta$
	n: power law index	
Tensorial	$\underline{\underline{\tau}}$: shear stress tensor	$\mathrm{II}_{\dot{\gamma}}$: second invariant of the
	$\underline{\underline{\dot{\gamma}}}$: deformation tensor	deformation tensor
	η, μ: viscosity	k: consistency
	A, B: constants	n: power law index
	λ: time constant (relaxation time)	

One-dimensional

Newton	$\tau = -\eta\,\dot{\gamma}$
Power law	$\tau = -k|\dot{\gamma}|^{n-1}\dot{\gamma}$
Bingham plastic	$\tau - \tau_0 = -\eta\dot{\gamma}$
Casson	$\sqrt{\tau} - \sqrt{\tau_0} = \sqrt{\eta|\dot{\gamma}|}$
Prandtl	$\tau = A \arcsin\left(\dfrac{\dot{\gamma}}{B}\right)$
Prandtl-Eyring	$\tau = -A \arcsin(B\dot{\gamma})$
Eyring	$\tau = \dfrac{\dot{\gamma}}{B} + c\sin\left(\dfrac{\tau}{A}\right)$
Powell-Eyring	$\tau = A\dot{\gamma} + B\mathrm{arcsinh}(C\dot{\gamma})$
Williamson	$\tau = \dfrac{A\dot{\gamma}}{B + \dot{\gamma}} + \eta_\infty\dot{\gamma}$

Tensorial

Newton	$\underline{\underline{\tau}} = \eta\,\underline{\underline{\dot{\gamma}}}$
Second-order	$\underline{\underline{\tau}} = A\underline{\underline{\dot{\gamma}}} + B\underline{\underline{\dot{\gamma}}}^2$
Power law	$\underline{\underline{\tau}} = k\left(\dfrac{1}{2}\mathrm{II}_{\dot{\gamma}}\right)^{\frac{n-1}{2}}\underline{\underline{\dot{\gamma}}}$ with $\mathrm{II}_{\dot{\gamma}} = \sum\limits_i \sum\limits_j \dot{\gamma}_{ij}\dot{\gamma}_{ij}$

Classical Maxwell $\underline{\underline{\tau}} + \lambda \dfrac{\partial \underline{\underline{\tau}}}{\partial t} = \eta_0 \underline{\underline{\dot{\gamma}}}$

(This equation is mathematically not objective)

Oldroyd-Maxwell fluid $\underline{\underline{\tau}} + \lambda \dfrac{\delta \underline{\underline{\tau}}}{\delta t} = \eta_0 \underline{\underline{\dot{\gamma}}}$ with

$$\frac{\delta \tau_{ij}}{\delta t} = \frac{\partial \tau_{ij}}{\partial t} + \sum_k \left\{ v_k \frac{\partial \tau_{ij}}{\partial x_k} - \tau_{kj} \frac{\partial v_i}{\partial x_k} - \tau_{ik} \frac{\partial v_j}{\partial x_k} \right\}$$

Jaumann-Maxwell fluid $\underline{\underline{\tau}} + \lambda \dfrac{\mathcal{D}\underline{\underline{\tau}}}{\mathcal{D}t} = \eta_0 \underline{\underline{\dot{\gamma}}}$, for $\dfrac{\mathcal{D}\underline{\underline{\tau}}}{\mathcal{D}t}$, see page 46.

Linear and logarithmic rheograms for recognition of rheological models

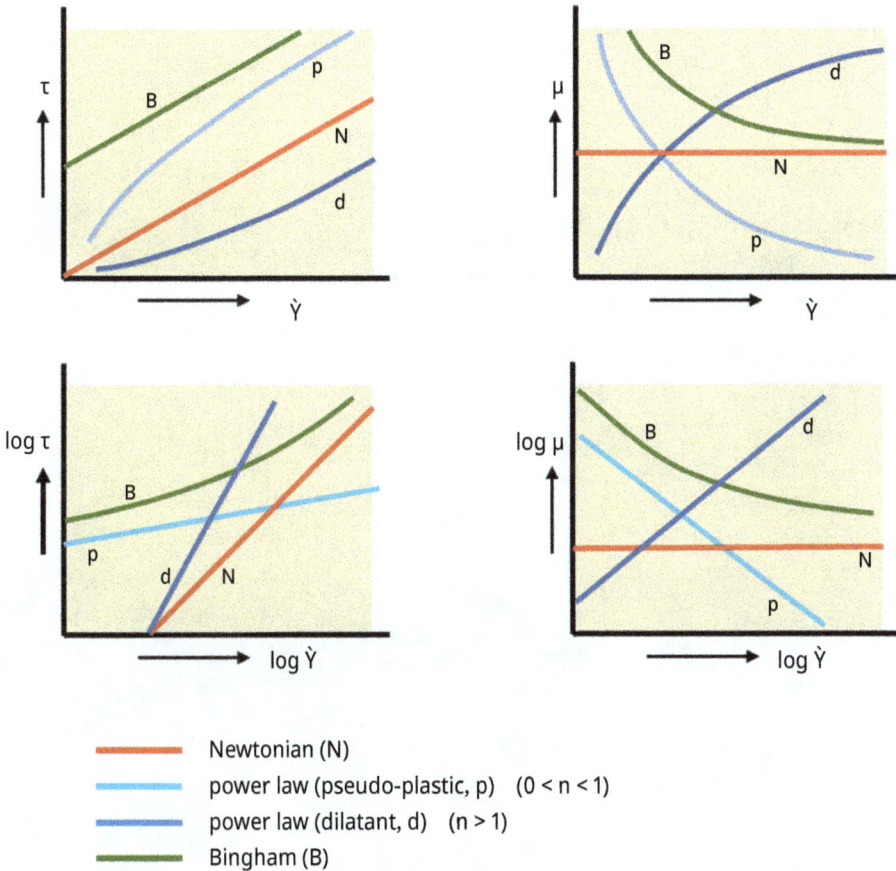

Newtonian (N)

power law (pseudo-plastic, p) $(0 < n < 1)$

power law (dilatant, d) $(n > 1)$

Bingham (B)

Fig. 4.22: Linear and logarithmic rheograms.

Pressure-throughput characteristics for laminar flow of liquids in a straight round tube

Notation

φ_v: volumetric throughput τ_w: shear stress at the wall

τ_0: yield stress R: tube radius

η: viscosity p: pressure

k: consistency τ: shear stress

n: power-law index

$$\dot{\gamma} = \frac{dv_z}{dr}$$

Newton $\tau = -\eta\dot{\gamma}$

$$\varphi_v = -\frac{\pi R^4}{8\eta}\frac{dp}{dz}$$

Power law $\tau = -k|\dot{\gamma}|^{n-1}\dot{\gamma}$

$$\varphi_v = \frac{n}{3n+1}\pi\,R^3\left|\frac{R}{2k}\frac{dP}{dz}\right|^{\frac{1}{n}}$$

Bingham plastic $\tau - \tau_0 = -\eta\dot{\gamma}$ $\quad(\tau > \tau_0)$

$$\varphi_v = -\frac{\pi R^3}{\eta}\left(\frac{\tau_0^4}{12} - \frac{R}{8}\frac{dP}{dz} - \frac{\tau_0}{3}\right)$$

Residence time distribution

Notation

E: "exit age" distribution V: volume
F: "internal age" distribution φ_v: volumetric throughput
t: time t_0: breakthrough time
τ: mean residence time Pe: Peclet number $\left(\text{Pe} = \frac{\langle v \rangle D}{\mathbb{D}}\right)$
ϑ: dimensionless time (= t/τ) n: number of ideally mixed tanks
σ: standard deviation

Mean residence time

$$\tau \equiv V/\varphi_v$$

$$\tau = \int_0^\infty t\, E(t)\; dt$$

$$\tau = \int_t^\infty (1 - F)\, dt$$

Variance of the distribution

$$\sigma^2 = \int_0^\infty (\vartheta - 1)^2\, E(\vartheta)\; d\vartheta$$

Relations

$$F(t) \quad = \int_0^t E(t)dt$$

$$1 - F(t) = \int_0^\infty E(t)\; dt$$

$$\int_0^\infty E(\vartheta)\; d\vartheta \quad = 1$$

$$\int_0^\infty 1 - F(\vartheta)d\vartheta = 1$$

Input signals

$t = 0$ $E = \infty$

$t \neq 0$ $E = 0$

$t < 0$ $F = 0$

$t > 0$ $F = 1$

Responses of systems

Plug flow

$E = 0$ $t \neq \tau$

$E = \infty$ $t = \tau$

$F = 0$ $t < \tau$

$F = 1$ $t > \tau$

$t_0 = \tau$

Ideally mixed tank

$E = e^{-\vartheta}$ $t > 0$

$F = 1 - e^{-\vartheta}$ $t > 0$

$t_0 = 0$

Laminar flow in a round tube

$E = \frac{1}{8\vartheta^3}$ $t > \tau/2$

$E = 0$ $t < \tau/2$

$F = 1 - \dfrac{1}{4\vartheta^2}$ $t > \tau/2$

$F = 0$ $t < \tau/2$

$t_0 = \tau/2$

Plug flow with axial dispersion

$$E = \tfrac{1}{2}\sqrt{\frac{\mathrm{Pe}}{\pi\vartheta}}\, \exp\left\{ -\frac{\mathrm{Pe}}{4\vartheta}(1-\vartheta)^2 \right\}$$

$$E \approx \tfrac{1}{2}\sqrt{\frac{\mathrm{Pe}}{\pi}}\, \exp\left\{ -\frac{\mathrm{Pe}}{4}(1-\vartheta)^2 \right\} (\mathrm{Pe} \gg 100)$$

$$F \approx \tfrac{1}{2}\left\{ 1 - \mathrm{erf}\left(\tfrac{1}{2}\frac{(1-\vartheta)}{\sqrt{\vartheta}}\sqrt{\mathrm{Pe}} \right) \right\} (\mathrm{Pe} \gg 100)$$

$$\sigma^2 = \frac{2}{\mathrm{Pe}} - \frac{2}{(\mathrm{Pe})^2}\left\{ 1 - \exp(-\mathrm{Pe}) \right\}$$

$$\sigma^2 \approx \frac{2}{\mathrm{Pe}} \;(\mathrm{Pe} \gg 100)$$

n ideally mixed tanks in series

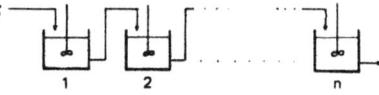

$$E = \frac{n^n \vartheta^{n-1}}{(n-1)!} e^{-n\vartheta}$$

$$E \approx \sqrt{\frac{n}{2\pi}} \vartheta^{n-1} \exp\{-n(\vartheta-1)\} \quad (n>5)$$

$$F \approx 1 - e^{-\vartheta} \left\{ 1 + n\vartheta + \frac{(n\vartheta)^2}{2!} + \frac{(n\vartheta)^3}{3!} + \ldots + \frac{(n\vartheta)^{n-1}}{(n-1)!} \right\} (n>5)$$

$$\sigma^2 = \frac{1}{n}$$

Quantities and their dimensions

Symbol	Dimension	Basic dimension	Quantity	Occurs in:
a	$\dfrac{m^2}{s}$	$\dfrac{m^2}{s}$	Thermal diffusivity	Fo, Gz, Le, Pe, Pr (see also λ)
C_P	$\dfrac{J}{kg\,K}$	$\dfrac{m^2}{s^2\,K}$	Specific heat	Da IV, St
d	m	m	Diameter, characteristic length	Ar, Bm, Bd, Da I, Da II, Da III, Da IV, Da V, De, El, Eo, f, Fo, Ga, Gz, Gr, Pe, Ra, Re, Sh, T, We, Ws
D	m	m	Stirrer diameter	Fr, P, Re
D_s	m	m	Spiral diameter	De
$\mathbb{D}$	$\dfrac{m^2}{s}$	$\dfrac{m^2}{s}$	Diffusion coefficient	Bi, Bo, Da III, Fo, Ha, Le, Pe, Sc, Sh, T
E	$\dfrac{J}{kg}$	$\dfrac{m^2}{s^2}$	Activation energy	Ah
g	$\dfrac{m}{s^2}$	$\dfrac{m}{s^2}$	Gravitational acceleration	Ar, Bd, Eo, Fr, Ga, Gr, Mo, Ri
h	$\dfrac{W}{m^2K}$	$\dfrac{kg}{Ks^3}$	Heat transfer coefficient	Bi, Nu, St
H	m	m	Height	Fr, Ri
k	$\dfrac{m}{s}$	$\dfrac{m}{s}$	Mass transfer coefficient	Bi, Ha, Sh
k_r	s^{-1}	$\dfrac{1}{s}$	Reaction constant	Da I, Da II, Da III, Da IV, Ha, T
L	m	m	Length (in transport direction)	Bo, Db, f, Gz
N	s^{-1}	$\dfrac{1}{s}$	Rotation rate	Fr, Po, Re, We
p	Pa	$\dfrac{kg}{m\,s^2}$	Pressure	f
P	W	$\dfrac{kg\,m^2}{s^3}$	Power added	Po
R	$\dfrac{J}{kgK}$	$\dfrac{m^2}{s^2\,K}$	Gas constant	Ah
t	s	s	Time	Fo
T	K	K	Temperature	Ah, Br, Da III, Da IV, Gr

(continued)

Symbol	Dimension	Basic dimension	Quantity	Occurs in:
v	$\dfrac{m}{s}$	$\dfrac{m}{s}$	Velocity	Bm, Bo, Br, C, Da I, Da IV, Da V, De, Db, f, Fr, Gz, Ho, Pe, Ra, Re, Ri, St, We, Ws
γ	K^{-1}	$\dfrac{1}{K}$	Bulk expansion coefficient	Gr
ϵ	Pa	$\dfrac{kg}{m\,s^2}$	Modulus of elasticity	C, Ho
η	Pa · s	$\dfrac{kg}{m\,s}$	Dynamic viscosity	Ar, Bm, Br, Da V, De, El, Ga, Mo, Re (see also v)
ϑ	s	s	Relaxation time	Db, El, Ws
λ	$\dfrac{W}{mK}$	$\dfrac{kg\,m}{s^3K}$	Thermal conductivity	Bi, Br, Da IV, Nu (see also a)
v	$\dfrac{m^2}{s}$	$\dfrac{m^2}{s}$	Kinematic viscosity	Gr, Pr, Sc (see also η)
ρ	$\dfrac{kg}{m^3}$	$\dfrac{kg}{m^3}$	Density, specific mass	Ar, C, Da III, Da V, De, El, f, Ga, Ho, Mo, Po, Ra, Re, Ri, St, We
$\Delta\rho$	$\dfrac{kg}{m^3}$	$\dfrac{kg}{m^3}$	Density difference	Ar, Bd, Eo, Mo, Ri
σ	$\dfrac{N}{m}$	$\dfrac{kg}{s^2}$	Surface tension	Bd, Eo, Mo, Ra, We
τ_0	Pa	$\dfrac{kg}{ms^2}$	Yield stress	Bm
ϑ_r	$\dfrac{J}{kg}$	$\dfrac{m^2}{s^2}$	Heat of reaction per unit mass	Da III, Da IV

Alphabetical list of dimensionless numbers

Symbol	Name	Formula	Physical significance	Relation to other dimensionless numbers	Used in
Ar	Archimedes	$\dfrac{gd^3\rho}{\eta^2}\Delta\rho$	$\dfrac{\text{Gravitational force}}{\text{Viscous force}}$	$Ar = Ga \times \dfrac{\Delta\rho}{\rho}$	Fluidization, free convection in viscous fluids
Ah	Arrhenius	$\dfrac{E}{RT}$	$\dfrac{\text{Activation energy}}{\text{Thermal energy}}$		Chemical reactions
Bm	Bingham	$\dfrac{\tau_0 d}{\eta v}$	$\dfrac{\text{Yield stress}}{\text{Viscous stress}}$		Bingham fluids
Bi	Biot (heat)	$\dfrac{h_1 L}{\lambda_2}$	$\dfrac{\text{External heat transfer resistance}}{\text{Internal heat transfer resistance}}$	Nusselt number for different media	Heat transfer between different media
Bi	Biot (mass)	$\dfrac{k_1 d}{\mathbb{D}_2}$	$\dfrac{\text{External mass transfer resistance}}{\text{Internal mass transfer resistance}}$	Sherwood number for different media	Mass transfer between different media
Bo	Bodenstein	$\dfrac{vL}{\mathbb{D}_{ax}}$	$\dfrac{\text{Convection}}{\text{Axial diffusion}}$	$Bo = Pe$	Diffusion in reactors
Bd	Bond	$\dfrac{\Delta\rho d^2 g}{\sigma}$	$\dfrac{\text{Gravitational force}}{\text{Surface tension force}}$	$Bd = Eo = \dfrac{We}{Fr}$	Bubbles and drops
Br	Brinkman	$\dfrac{\eta v^2}{\lambda\Delta T}$	$\dfrac{\text{Viscous dissipation}}{\text{Conduction of heat}}$		Flow of viscous liquids
C	Cauchy	$\dfrac{\rho v^2}{\varepsilon}$	$\dfrac{\text{Inertia force}}{\text{Compressibility force}}$	$C = Ho$	Flow of compressible media
Da I	Damköhler I	$\dfrac{k_r d}{v}$	$\dfrac{\text{Chemical reaction rate}}{\text{Convective mass transfer rate}}$		Chemical reactions

Da II	Damköhler II	$\dfrac{k_r d^2}{\mathbb{D}}$	$\dfrac{\text{Chemical reaction rate}}{\text{Diffusive mass transfer rate}}$		Chemical reactions
Da III	Damköhler III	$\dfrac{\varphi_r k_r d}{\rho c_p v T}$	$\dfrac{\text{Heat of reaction}}{\text{Convective transport of heat}}$		Chemical reactions
Da IV	Damköhler IV	$\dfrac{\varphi_r k_r d^2}{\lambda T}$	$\dfrac{\text{Heat of reaction}}{\text{Conductive transport of heat}}$		Chemical reactions
Da V	Damköhler V	$\dfrac{\rho v d}{\eta}$	$\dfrac{\text{Inertia forces}}{\text{Viscous forces}}$	$\text{Da V} = \text{Re}$	Chemical reactions
De	Dean	$\dfrac{v d \rho}{\eta}\sqrt{\dfrac{d}{D_s}}$	$\dfrac{\text{Centrifugal forces}}{\text{Viscous forces}}$	$\text{De} = \text{Re}\sqrt{\dfrac{d}{D_s}}$	Flow in curved channels
Db	Deborah	$\dfrac{\vartheta v}{L}$	$\dfrac{\text{Relaxation time}}{\text{Characteristic process time}}$		Flow of viscoelastic media
El	Elasticity	$\dfrac{\vartheta \eta}{\rho d^2}$	$\dfrac{\text{Elastic forces}}{\text{Inertia forces}}$	$\text{El} = \dfrac{\text{Ws}}{\text{Re}}$	Flow of viscoelastic media
Eo	Eötvös	$\dfrac{d^2 g}{\sigma}\Delta\rho$	$\dfrac{\text{Gravity forces}}{\text{Surface tension forces}}$	$\text{Eo} = \dfrac{\text{We}}{\text{Fr}} = \text{Bd}$	Bubbles and droplets
f	Fanning friction factor	$\dfrac{d \Delta p}{2 \rho v^2 L}$	$\dfrac{\text{Shear stress energy at the wall}}{\text{Kinetic energy}}$		Flow through tubes and channels
Fo	Fourier (mass)	$\dfrac{\mathbb{D} t}{d^2}$	$\dfrac{\text{Process time}}{\text{Effective diffusion time}}$	$\text{Fo}_s = \dfrac{\text{Fo}_w}{\text{Le}}$	Diffusion
Fo	Fourier (heat)	$\dfrac{a t}{d^2}$	$\dfrac{\text{Process time}}{\text{Effective time for heat conduction}}$	$\text{Fo}_w = \text{Fo}_s \times \text{Le}$	Heat conduction

(continued)

(continued)

Symbol	Name	Formula	Physical significance	Relation to other dimensionless numbers	Used in
Fr	Froude	$\dfrac{v^2}{gH}$	$\dfrac{\text{Inertia forces}}{\text{Gravity forces}}$		Waves and surfaces
Fr	Froude (rotation)	$\dfrac{DN^2}{g}$	$\dfrac{\text{Inertia forces}}{\text{Gravity forces}}$		Mixing with free surfaces
Ga	Galileo	$\dfrac{d^3 g \rho^2}{\eta^2}$	$\dfrac{\text{Bouyancy forces}}{\text{Viscous forces}}$	$Ga = Ar \times \dfrac{\rho}{\Delta\rho}$	Gravitational flow of viscous fluids, free convection
Gz	Graetz	$\dfrac{aL}{d^2 v}$	$\dfrac{\text{Conductive heat transfer}}{\text{Convective heat transfer}}$		Heat transfer to flowing media
Gr	Grashof	$\dfrac{d^3 g}{v^2}\gamma\Delta T$	$\dfrac{\text{Bouyancy forces}}{\text{Viscous forces}}$		Free convection
Ha	Hatta	$\dfrac{\sqrt{\mathbb{D}k_r}}{k}$	$\dfrac{\text{Mass transfer with chemical reaction}}{\text{Mass transfer without chemical reaction}}$		Mass transfer with chemical reaction
Ho	Hooke	$\dfrac{\rho v^2}{\epsilon}$	$\dfrac{\text{Inertia forces}}{\text{Elastic forces}}$	$Ho = C$	Flow of viscoelastic media
Le	Lewis	$\dfrac{a}{\mathbb{D}}$	$\dfrac{\text{Thermal boundary layer thickness}}{\text{Mass transfer boundary layer thickness}}$	$Le = \dfrac{Sc}{Pr}$	Combined heat and mass transfer
Mo	Morton	$\dfrac{g\eta^4\Delta\rho}{\rho^2\sigma^3}$	Physical properties of phases		Two-phase flow
Nu	Nusselt	$\dfrac{hd}{\lambda}$	$\dfrac{\text{Total heat transfer}}{\text{Conductive heat transfer}}$	$Nu = Pe \times St$ $= Re \times Pr \times St$	Heat transfer to flowing media

	Name	Formula	Ratio	Relations	Application
Pe	Péclet (heat)	$\dfrac{vd}{a}$	$\dfrac{\text{Convective heat transfer}}{\text{Conductive heat transfer}}$	$\begin{aligned}\text{Pe} &= \text{Re} \times \text{Pr}\\ &= \text{Pe} \times \text{Le}\end{aligned}$	Heat transfer to flowing media
Pe	Péclet (mass)	$\dfrac{vd}{\mathbb{D}}$	$\dfrac{\text{Convective mass transfer}}{\text{Diffusive mass transfer}}$	$\begin{aligned}\text{Pe} &= \text{Re} \times \text{Sc}\\ &= \text{Pe}/\text{Le}\end{aligned}$	Mass transfer to flowing media
Po	Power number	$\dfrac{P}{\rho N^3 D^5}$	$\dfrac{\text{Power added}}{\text{Power transferred to kinetic energy}}$		Stirrers and pumps
Pr	Prandtl	$\dfrac{\nu}{a}$	$\dfrac{\text{Hydrodynamic boundary layer thickness}}{\text{Thermal boundary layer thickness}}$	$\text{Pr} = \dfrac{\text{Sc}}{\text{Le}} = \dfrac{\text{Pe}}{\text{Re}}$	Heat transfer to flowing media
Ra	Rayleigh	$\nu\sqrt{\dfrac{\rho d}{\sigma}}$	$\dfrac{\text{Inertia forces}}{\text{Surface tension forces}}$	$\text{Ra} = \sqrt{\text{We}}$	Break up of jets and droplets
Re	Reynolds	$\dfrac{\rho v d}{\eta}$	$\dfrac{\text{Inertia forces}}{\text{Viscous forces}}$		Flow
Re	Reynolds (rotation)	$\dfrac{D^2 N \rho}{\eta}$	$\dfrac{\text{Inertia forces}}{\text{Viscous forces}}$		Mixing
Ri	Richardson	$\dfrac{gH\Delta\rho}{v^2 \rho}$	$\dfrac{\text{Potential energy}}{\text{Kinetic energy}}$	$\text{Ri} = \dfrac{1}{\text{Fr}}\dfrac{\rho-\rho_0}{\rho}$	Mixing of stratified layers
Sc	Schmidt	$\dfrac{\nu}{\mathbb{D}}$	$\dfrac{\text{Hydrodynamic boundary layer}}{\text{Mass transfer boundary layer}}$	$\begin{aligned}\text{Sc} &= \text{Pr} \times \text{Le}\\ &= \text{Pe}/\text{Re}\end{aligned}$	Mass transfer to flowing media
Sh	Sherwood	$\dfrac{kd}{\mathbb{D}}$	$\dfrac{\text{Total mass transfer}}{\text{Diffusive mass transfer}}$		Mass transfer to flowing media
St	Stanton	$\dfrac{h}{\rho C_p v}$	$\dfrac{\text{Heat transferred}}{\text{Heat capacity of the fluid}}$	$\text{St} = \dfrac{\text{Nu}}{\text{Re} \times \text{Pr}} = \dfrac{\text{Nu}}{\text{Pe}}$	Heating of flowing media
T	Thiele modulus	$d\sqrt{\dfrac{k_r}{\mathbb{D}}}$	$\dfrac{\text{Chemical reaction rate}}{\text{Diffusive mass transfer rate}}$	$T = \sqrt{\text{DaII}}$	Porous catalysts

(continued)

(continued)

Symbol	Name	Formula	Physical significance	Relation to other dimensionless numbers	Used in
We	Weber	$\dfrac{\rho v^2 d}{\sigma}$	$\dfrac{\text{Inertia forces}}{\text{Surface tension forces}}$	$We = Ra^2 = Eo \times Fr$	Bubbles and droplets
Ws	Weissenberg	$\dfrac{\vartheta v}{d}$	$\dfrac{\text{Viscoelastic forces}}{\text{Viscous forces}}$	$Ws = El \times Re$	Flow of viscoelastic media

Frequently used dimensionless correlations

Flow

Pipe flow
Pressure drop

$$\Delta p = 4f \cdot \frac{1}{2}\, \rho \langle v \rangle^2\, \frac{L}{d}$$

Friction coefficients:

Laminar flow in tubes (Re < 1,000)
$$4f = \frac{64}{Re}$$
(Hagen-Poiseuille)

Turbulent flow in smooth tubes: (4,000 < Re < 10^5) $4f = 0.316\, Re^{-\frac{1}{4}}$ (Blasius)

Flow resistance of an object

$$F = C_d\, A_\perp\, \frac{1}{2}\, \rho\, v^2$$

For C_d values, see page 103 and 104.

Heat transfer

Flow in tubes
Laminar

Entrance region:	$Nu = 1.08 Gz^{-\frac{1}{3}}$	Gz < 0.05
	$\langle Nu \rangle = 1.62 Gz^{-\frac{1}{3}}$	Gz < 0.05
Developed profile:	$Nu = \langle Nu \rangle = 3.66$	Gz > 0.1
Turbulent	$\langle Nu \rangle = 0.027 Re^{0.8}\, Pr^{0.33}$	Re > 10^4; Pr ≥ 0.7

Flow around obstacles

Plate (// flow):	$\langle Nu \rangle = 0.332\, Re^{0.5}\, Pr^{0.33}$	Re < 3 × 10^5; Pr ≥ 0.7
Long cylinders (⊥ flow):	$\langle Nu \rangle = 0.57\, Re^{0.5}\, Pr^{0.33}$	10 < Re < 10^4; Pr ≥ 0.7; Pe ≥ 1
Spheres	$\langle Nu \rangle = 2.0 + 0.66\, Re^{0.50}\, Pr^{0.33}$	10 < Re < 10^4; Pr ≥ 0.7; Pe ≥ 1
Packed beds:	$\langle Nu \rangle = (1.8 \pm 0.3)\, Re_0^{0.5}\, Pr^{0.33}$	30 < Re_0 < 3 × 10^3; Pr ≥ 1
		Re_0: Re for empty bed

Internal circulation in bubbles and drops

Bubbles and drops are rigid (no internal circulation) if: $d < \sqrt{\frac{\sigma}{g\,\Delta\rho}}$ or $\mathrm{Bd} < 1$

Heat transfer from bubbles and drops, resistance in the continuous phase (Bi ≪ 1)

With internal circulation:

Re < 1
$$\mathrm{Nu_c} = 0.65\left(\frac{\mu_c}{\mu_c + \mu_d} \times \mathrm{RePr}\right)^{\frac{1}{2}}$$

1 < Re < 1,500
$$\mathrm{Nu_c} = 5.52\left(\frac{\mu_c + \mu_d}{2\mu_c + 3\mu_d} \times \mathrm{RePr}\right)^{3.47}\left(\frac{d\sigma\rho_c}{\mu_c^2}\right)^{0.056}\sqrt{\mathrm{Re}\cdot\mathrm{Pr}}$$

Without internal circulation:

$10 < \mathrm{Re} < 10^4\ \mathrm{Pr} \geq 0.7$
$$\langle\mathrm{Nu}\rangle = 2.0 + 0.66\,\mathrm{Re}^{0.50}\,\mathrm{Pr}^{0.33}$$

Heat transfer from bubbles and drops, resistance in the dispersed phase (Bi ≫1)

With internal circulation:

$$\mathrm{Nu_d} = 0.00375\,\mathrm{Re}\cdot\mathrm{Pr}\left(\frac{\mu_c}{\mu_c + \mu_d}\right)$$

Without internal circulation:

Fo > 0.1 $\mathrm{Nu_d} = 6.6$

Free convection

Vertical plate: $\langle\mathrm{Nu}\rangle = 0.55\,(\mathrm{Gr}\times\mathrm{Pr})^{0.25}$ $10^4 < \mathrm{Gr}\times\mathrm{Pr} < 10^8$ (laminar flow)

$\langle\mathrm{Nu}\rangle = 0.13\,(\mathrm{Gr}\times\mathrm{Pr})^{0.33}$ $\mathrm{Gr}\times\mathrm{Pr} > 10^8$ (turbulent flow)

Two horizontal plates: $T_{\mathrm{lower}} > T_{\mathrm{upper}}$

$\langle\mathrm{Nu}\rangle = 1$ $\mathrm{Gr}\times\mathrm{Pr} < 10^3$ (conduction)

$\langle\mathrm{Nu}\rangle = 0.15\,(\mathrm{Gr}\times\mathrm{Pr})^{0.25}$ $10^4 < \mathrm{Gr}\times\mathrm{Pr} < 10^7$ (occurrence of Benard cells)

$\langle\mathrm{Nu}\rangle = 0.17\,(\mathrm{Gr}\times\mathrm{Pr})^{0.33}$ $\mathrm{Gr}\times\mathrm{Pr} > 10^7$ (turbulent flow)

Two horizontal plates: $T_{\text{lower}} < T_{\text{upper}}$

$\langle\text{Nu}\rangle = 1$ for all $\text{Re} \times \text{Pr}$

Instationary heat transfer

Short time: $\text{Nu} = 0.57\,\text{Fo}^{-\frac{1}{2}}$ $\text{Fo} \ll 0.1$

               $\langle\text{Nu}\rangle = 1.14\,\text{Fo}^{-\frac{1}{2}}$ $\text{Fo} \ll 0.1$

Long times: $\text{Nu} = \langle\text{Nu}\rangle = 4.93$ plate $\text{Fo} > 1$

               $\text{Nu} = \langle\text{Nu}\rangle = 6.6$ sphere $\text{Fo} > 1$

               $\text{Nu} = \langle\text{Nu}\rangle = 5.8$ cylinder $\text{Fo} > 1$ $L \gg d$

Mass transfer

Flow in tubes
Laminar:

Entrance region: $\text{Sh} = 1.08\text{Gz}^{-\frac{1}{3}}$ $\text{Gz} < 0.05$

                   $\langle\text{Sh}\rangle = 1.62\text{Gz}^{-\frac{1}{3}}$ $\text{Gz} < 0.05$

Developed profile: $\text{Sh} = \langle\text{Sh}\rangle = 3.66$ $\text{Gz} > 0.1$

Turbulent:

For all Gz: $\langle\text{Sh}\rangle = 0.027\text{Re}^{0.8}\,\text{Sc}^{0.33}$ $\text{Re} > 10^4;\ \text{Sc} \geq 0.7$

Flow around obstacles

Plate // flow: $\langle\text{Sh}\rangle = 0.332\,Re^{0.5}\,Sc^{0.33}$ $\text{Re} < 3 \cdot 10^5;\ \text{Sc} \geq 0.7$

Long cylinders ⊥ flow $\langle\text{Sh}\rangle = 0.57\,Re^{0.5}\,Sc^{0.33}$ $10 < \text{Re} < 10^4;\ \text{Sc} \geq 0.7:\ \text{Pe} \geq 1$

Spheres $\langle\text{Sh}\rangle = 2.0 + 0.66\,Re^{0.50}\,Sc^{0.33}$ $10 < \text{Re} < 10^4;\ \text{Sc} \geq 0.7:\ \text{Pe} \geq 1$

Packed beds: $\langle\text{Sh}\rangle = (1.8 \pm 0.3)\,Re^{0.5}\,Sc^{0.33}$ $30 < \text{Re} < 3 \times 10^3;\ \text{Sc} \geq 1$

Mass transfer from bubbles and drops, resistance in the continuous phase (Bi ≪1)
With internal circulation:

$\text{Re} < 1$ $\text{Sh}_{\text{c}} = 0.65 \left(\dfrac{\mu_{\text{c}}}{\mu_{\text{c}} + \mu_{\text{d}}} \times \text{ReSc} \right)^{\frac{1}{2}}$

$1 < \text{Re} < 1500 \quad Sh_c = 5.52 \left(\dfrac{\mu_c + \mu_d}{2\mu_c + 3\mu_d} \times ReSc \right)^{3.47} \left(\dfrac{d\sigma\rho_c}{\mu_c^2} \right)^{0.056} \sqrt{Re \cdot Sc}$

Without internal circulation:

$10 < \text{Re} < 10^4 \ \text{Sc} \geq 0.7 \quad \langle \text{Sh} \rangle = 2.0 + 0.66 \, \text{Re}^{0.50} \, \text{Sc}^{0.33}$

Mass transfer from bubbles and drops, resistance in the continuous phase (Bi ≪1)
With internal circulation:

$$\text{Sh}_d = 0.00375 \, \text{Re} \cdot \text{Sc} \left(\dfrac{\mu_c}{\mu_c + \mu_d} \right)$$

Without internal circulation:

$\text{Fo} > 0.1 \qquad \text{Sh}_d = 6.6$

Free convection

Vertical plate:

$\langle \text{Sh} \rangle = 0.55 \, (\text{Gr} \times \text{Sc})^{0.25} \quad 10^4 < \text{Gr} \times \text{Sc} < 10^8 \ (\text{laminar flow})$
$\langle \text{Sh} \rangle = 0.13 \, (\text{Gr} \times \text{Sc})^{0.33} \quad \text{Gr} \times \text{Sc} > 10^8 \ (\text{turbulent flow})$

Two horizontal plates: $\rho_{\text{lower}} < \rho_{\text{upper}}$

$\langle \text{Sh} \rangle = 1 \qquad\qquad\qquad\quad \text{Gr} \times \text{Sc} < 10^3 \qquad (\text{conduction})$
$\langle \text{Sh} \rangle = 0.15 \, (\text{Gr} \times \text{Sc})^{0.25} \quad 10^4 < \text{Gr} \times \text{Sc} < 10^7 \ (\text{occurrence of Benard cells})$
$\langle \text{Sh} \rangle = 0.17 \, (\text{Gr} \times \text{Sc})^{0.33} \quad \text{Gr} \times \text{Sc} > 10^7 \qquad (\text{turbulent flow})$

Two horizontal plates: $\rho_{\text{lower}} > \rho_{\text{upper}}$

$\langle \text{Sh} \rangle = 1 \qquad \text{for all Gr} \times \text{Sc}$

Instationary mass transfer

Short time: $\text{Sh} = 0.57\,\text{Fo}^{-\frac{1}{2}}$ $\qquad \text{Fo} \ll 0.1$

$\qquad\qquad \langle \text{Sh} \rangle = 1.14\,\text{Fo}^{-\frac{1}{2}}$ $\qquad \text{Fo} \ll 0.1$

long times: $\text{Sh} = \langle \text{Sh} \rangle = 4.93$ plate $\quad \text{Fo} > 1$
$\qquad\qquad \text{Sh} = \langle \text{Sh} \rangle = 6.6$ sphere $\quad \text{Fo} > 1$
$\qquad\qquad \text{Sh} = \langle \text{Sh} \rangle = 5.8$ cylinder $\;\text{Fo} > 1\, L \gg d$

List of symbols and abbreviations

Symbols

A	Surface	m^2
C_d	Drag coefficient	–
f	Friction coefficient	–
F	Force	N
d	Diameter	m
g	Gravitational constant	m/s^2
L	Length	m
T	Temperature	K or °C
v	Velocity	m/s
μ	Viscosity	Pa s
ρ	Density	kg/m^3
σ	Surface tension	N/m

Dimensionless groups

Fo	Fourier
Gr	Grashof
Gz	Graetz
Nu	Nusselt
Pe	Péclet
Pr	Prandtl
Re	Reynolds
Sc	Schmidt
Sh	Sherwood
$< \cdots >$	Average value of . . .

Indices

d	Dispersed phase
c	Continuous phase

Analogies between heat transfer and mass transfer

Chilton-Colburn analogy (*J*-factor analogy)
Heat transfer number

$$J_h = \text{Nu } (\text{Re})^{-1}(\text{Pr})^{-\frac{1}{3}}$$

Mass transfer number

$$J_D = \text{Sh } (\text{Re})^{-1}(\text{Sc})^{-\frac{1}{3}}$$

For geometrical similarity

$$J_H = J_D$$

For turbulent flow through pipes and past flat plates

$$J_H = J_D = \frac{1}{2}f$$

Analogies

Heat transfer	Mass transfer
$\rho\, C_p\, T$	C_A
$\dfrac{\lambda}{\rho\, C_p} = a$	$\mathbb{D}_A$
Φ''_w	$\Phi''_{mol,A}$
$\dfrac{h}{\rho\, C_p}$	k
Nu	Sh
Pr	Sc

Drag coefficient C_w for flow around obstacles

Drag coefficients

Shape	Flow direction	Validity	C_d	
Flat plate		Laminar flow Re $< 10^7$	$1.33\,\mathrm{Re}^{-0.5}$ $0.074\,\mathrm{Re}^{-0.2}$	
		Re $> 10^3$	*L/d*	
			1	1.18
			5	1.2
			10	1.3
			20	1.5
			30	1.6
			∞	1.95
Disk		Re $> 10^3$	1.17	
Sphere		Re < 1 $10^3 < \mathrm{Re} < 3 \times 10^5$ $\mathrm{Re} > 3 \times 10^5$	24/Re 0.47 0.2	
Hollow semisphere		$10^4 < \mathrm{Re} < 10^6$	0.34	
		$10^4 < \mathrm{Re} < 10^6$	1.42	
Filled semisphere		$10^4 < \mathrm{Re} < 10^6$	1.17	
		$10^4 < \mathrm{Re} < 10^6$	0.42	
Cylinder		$10^3 < \mathrm{Re} < 10^5$	*L/d*	
			1	0.63
			5	0.8
			10	0.83
			20	0.93
			30	1.0
			∞	1.2

(continued)

Shape	Flow direction	Validity	C_d
Square rod		$Re = 3.5 \times 10^4$	2
Parachute			1.4
Cars	T-Fort		0.85
	Chevrolet (1950)		0.50
	Modern pickup truck		0.45
	Modern family car		0.27
	Sports car		0.25

Packed and fluidized beds

Packed beds

Hydraulic diameter: $d_h = \dfrac{2}{3}\dfrac{\varepsilon d_0}{(1-\varepsilon)}$

Reynolds number: $Re_h = \dfrac{2}{3}\dfrac{\rho v_0 d_0}{\mu(1-\varepsilon)}$

Friction losses: $A_w = \dfrac{3}{2}C_w \times \dfrac{1}{2}v_0^2 \dfrac{(1-\varepsilon)}{\varepsilon^3}\dfrac{L}{d_0}$

Friction coefficient: $C_w = \dfrac{150}{Re_h}$ if $Re_h < 10$ (Blake-Kozeny's equation)

$C_w = 2.3$ if $Re_h > 1,000$ (Burke-Plummer's equation)

$C_w = 2.3 + \dfrac{150}{Re_h}$ (Ergun's equation)

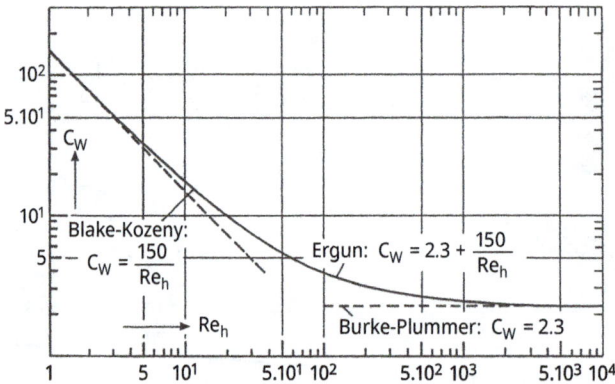

Fig. 4.23: Drag coefficient in a packed bed as function of Reynolds based on the hydraulic diameter.

Drag coefficient in a packed bed as function of Reynolds based on the hydraulic diameter

Fluidized beds

Pressure drop: $\Delta p = (1 - \varepsilon)(\rho_p - \rho_f)gH$

Low Reynolds number spherical particles (Re$_h$ < 20):
Minimum fluidization velocity:

$$v_{mf} = \frac{gd_p^2\left(\rho_p - \rho_f\right)}{150\mu_f}\left(\frac{\varepsilon_{mf}^3}{1 - \varepsilon_{mf}}\right)$$

Terminal velocity (blowout):

$$v_t = \frac{gd^2\left(\rho_p - \rho_f\right)}{18\mu_f}$$

High Reynolds number spherical particles (Re$_h$ > 1,000):
Minimum fluidization velocity:

$$v_{mf} = \sqrt{\frac{gd_p\left(\rho_p - \rho_f\right)}{1.75\,\rho_f}\,\varepsilon_{mf}^3}$$

Terminal velocity:

$$v_t = 1.75\sqrt{\frac{gd_p\left(\rho_p - \rho_f\right)}{\rho_p}}$$

Sedimentation velocity of a swarm of particles

$$v_{swarm} = v_s \cdot \varepsilon^n \quad \text{(Richardson and Zaki's equation*)}$$

where

0 < Re < 0.2	$n = 4.6 + 20\,d_p/d_0$
0.2 < Re < 1	$n = (4.4 + 18\,d_p/d_0)\,\text{Re}^{-0.03}$
1 < Re < 200	$n = (4.4 + 18\,d_p/d_0)\,\text{Re}^{-0.1}$
200 < Re < 500	$n = 4.4\,\text{Re}^{-0.1}$
Re > 500	$n = 2.39$

Quick reference:

Re'	< 0.1	1	10	100	>500
n	4.65	4.35	3.53	2.80	2.39

and $Re' = \dfrac{\rho v_s d_p}{\mu}$

*Richardson, J.F. and Zaki, W.N., Trans. Inst. Chem. Eng 32 (1954) 35–53

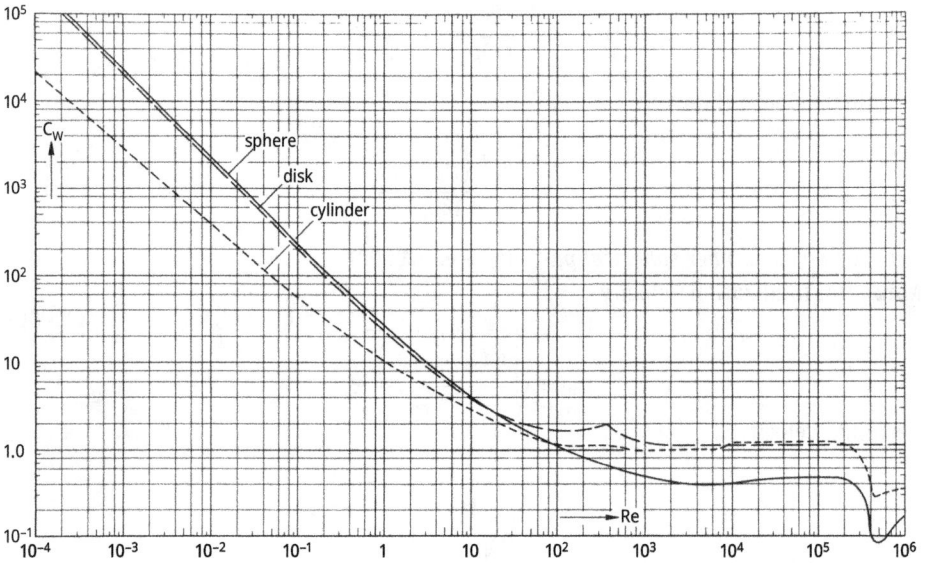

Fig. 4.24: Drag coefficient for flow around objects.

Symbol list

Symbols		Dimension	Indices	
A	Friction loss	m^2/s^2	h	Hydraulic
c	Friction coefficient	–	0	Empty bed
d	Diameter	m	w	Friction
g	Gravitational constant	m/s^2	p	Particle
L	Bed length	m	f	Fluidum
n	Constant	–	mf	Minimum fluidization
v	Velocity	m/s	t	Terminal (blowout)
ε	Void fraction	–	swarm	Swarm of particles
μ	Viscosity	Pa s	s	Single particle
ρ	Density	kg/m^3	ε	Void fraction

Friction coefficient K_w for flow through tube systems

Reynolds for downstream conditions $Re > 10^5$

Smooth bent							
			$K_w = \left\{ 0.163 \left(\dfrac{D}{R} \right)^{3.5} + 0.131 \right\} \dfrac{\alpha}{90}$				

Sharp bent		α	90	120	135	150	165
		K_w	1.30	0.50	0.26	0.11	0.02

Kink		α	40	60	80	90	100	120	140	160
		K_w	2.43	1.86	1.26	0.98	0.74	0.36	0.14	0.05

Sudden contraction	
	$K_w = 0.45 \left\{ 1 - \dfrac{A_2}{A_1} \right\}$

Gradual contraction	α	10	20	30	40	50	60	70	80
	K_w	0.16	0.20	0.24	0.28	0.31	0.32	0.34	0.35

Sudden enlargement	
	$K_w = \left\{ \dfrac{A_2}{A_1} - 1 \right\}^2$

Gradual enlargement	
	$K_w = K \left\{ 1 - \dfrac{A_1}{A_2} \right\}^2$

α	<10	10	20	30	40	50	60	70	80	90
K	0	0.17	0.41	0.71	0.90	1.03	1.12	1.13	1.10	1.05

Butterfly valve		α	20 ·	30	40	50	60	70	80	85
		K_w	751	118	33	11	4	1.5	0.52	0.24

Ball valve		α	25	30	40	50	60	70	80	85
		K_w	486	206	53	17	5.5	1.6	0.29	0.05

Gate valve			Open	¾	½	¼
		K_w	0.2	0.9	4.5	24

Glove valve		Open	¾	½	¼
	K_w	9	13	36	112

(continued)

T-junction			K_w	1.3	1.5	1.0	0.4
Bents	45°	Smooth	90°	Smooth	90°	Sharp	180°
K_w		0.35		0.75		1.3	1.5
Pipe entrance				Sharp			Rounded
K_w				0.5			0.05
Pipe exit				Free yet			Confined yet
K_w				0			1.0

Friction factor for flow in tubes

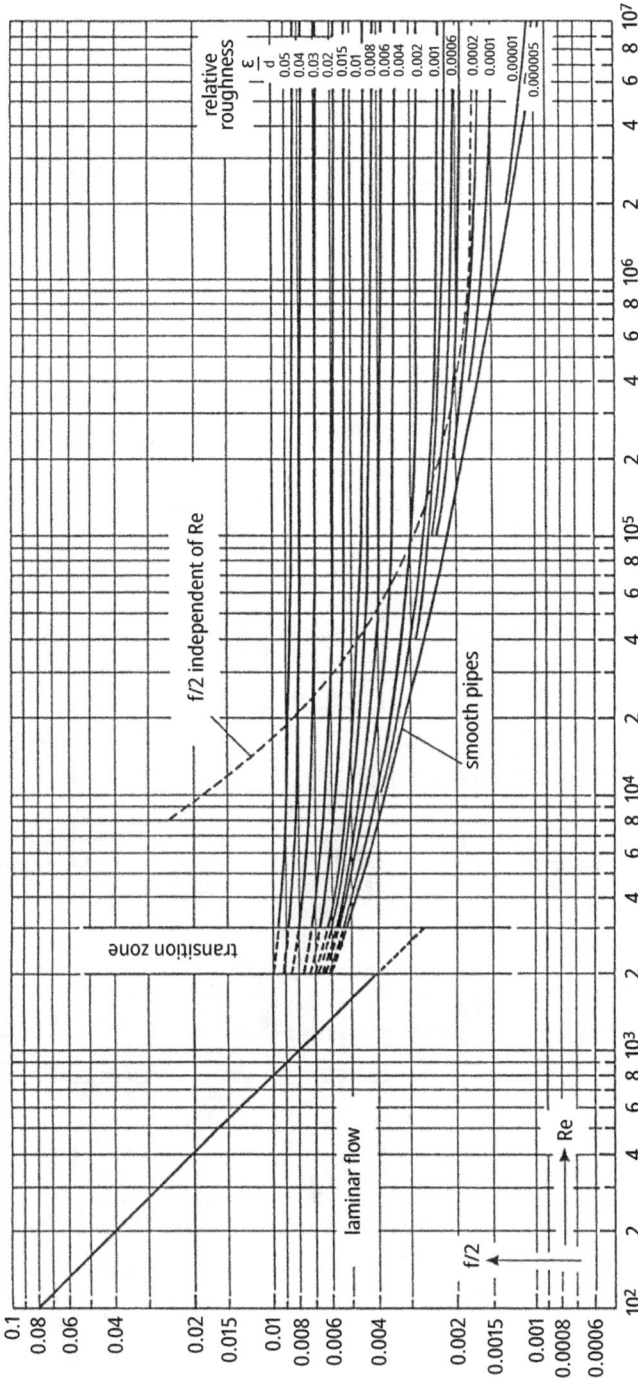

Fig. 4.25: Friction factor for flow in tubes.

Friction factors for pressure flow in pipes

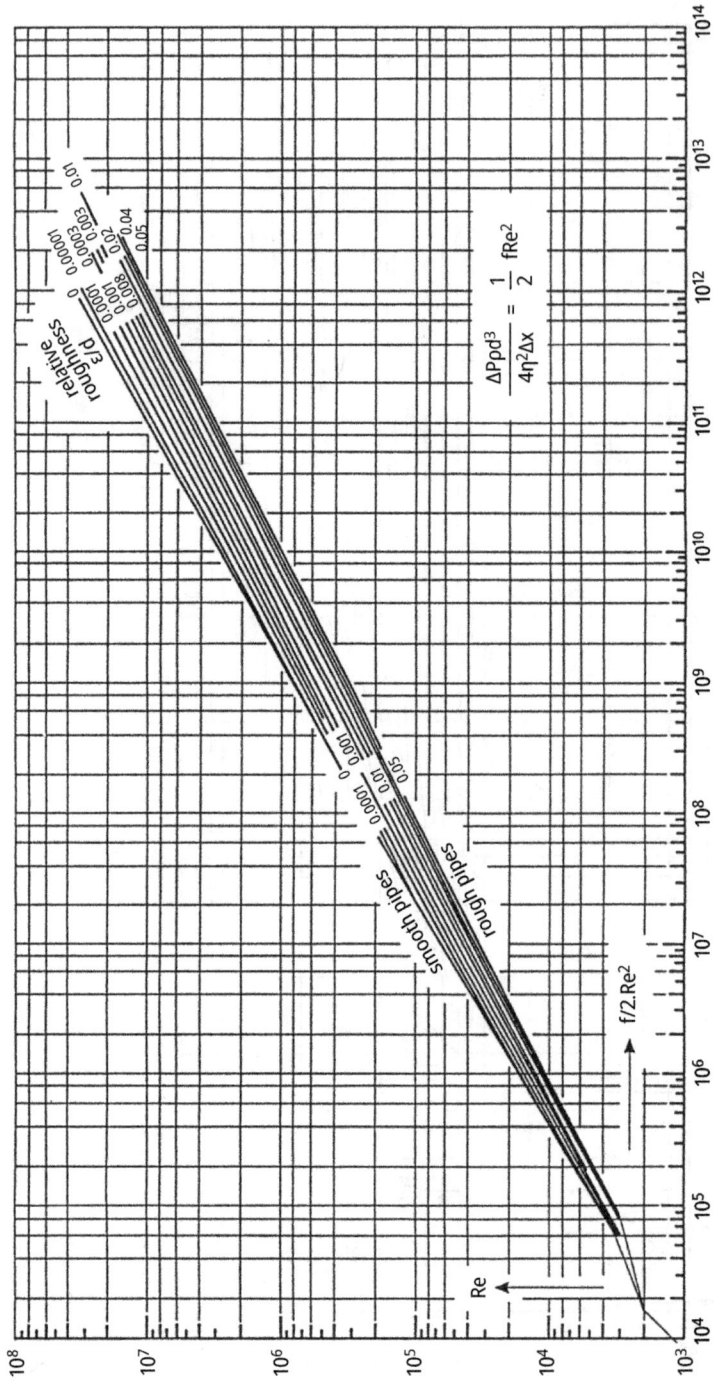

Fig. 4.26: Determination of Re and friction factors for known pressure drops in circular pipes.

Hydraulic diameters and Reynolds numbers

General	$d_h = \dfrac{4 \times \text{wet surface}}{\text{wetted circumference}}$		
Round tube:	$d_h = d$	$Re = \dfrac{\rho v d}{\mu}$	
Square tube:	$d_h = a$	$Re = \dfrac{\rho v a}{\mu}$	
Rectangular channel:	$d_h = \dfrac{4hw}{w+2h}$	$Re = 4\dfrac{\rho v}{\mu}\dfrac{hw}{w+2h}$	
Half round channel	$d_h = d$	$Re = \dfrac{\rho v d}{\mu}$	
Packed bed:	$d_h = \dfrac{2}{3}\dfrac{d_0}{1-\varepsilon}$ (ε: bed porosity)	$Re = \dfrac{2}{3}\dfrac{\rho v_0 d_0}{\mu(1-\varepsilon)}$	
Triangular channel:	$d_h = 2h\sin\dfrac{\alpha}{2}$	$Re = 2\dfrac{\rho v h}{\mu}\sin\dfrac{\alpha}{2}$	

The power number Po of impellers as a function of Re

(P = power input (watt), N = impeller speed (s^{-1}), D = impeller diameter (m), w = height of impeller blade).

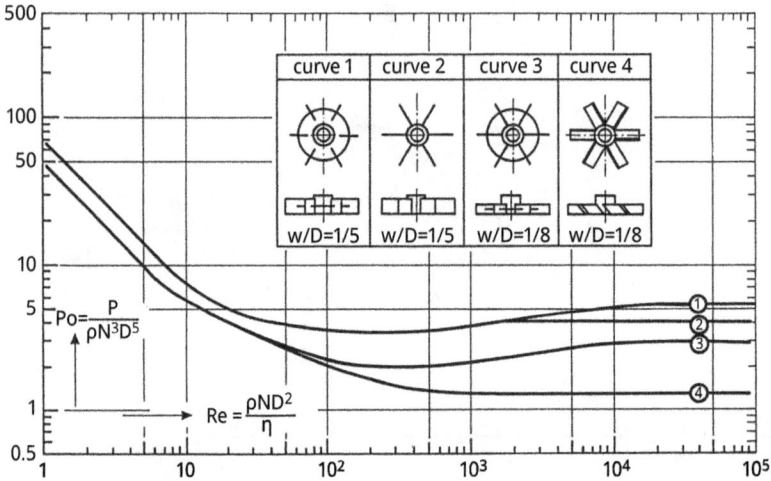

Fig. 4.27: Power number vs Reynolds number for different stirrers.

Humidity diagram for air-water systems at atmospheric pressure (I)

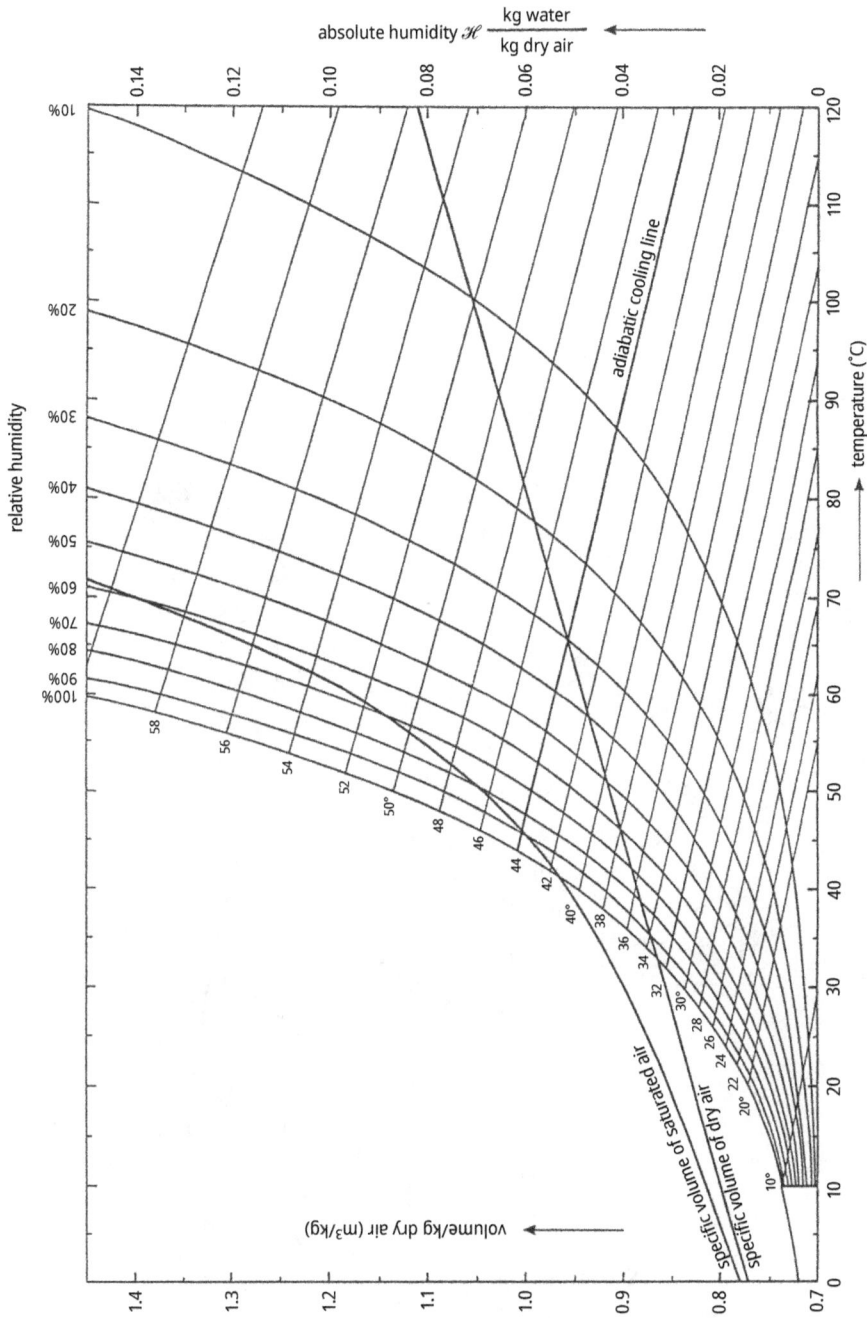

Fig. 4.28: Humidity diagram for air-water systems (high temperatures).

Humidity diagram for air-water systems at atmospheric pressure (II)

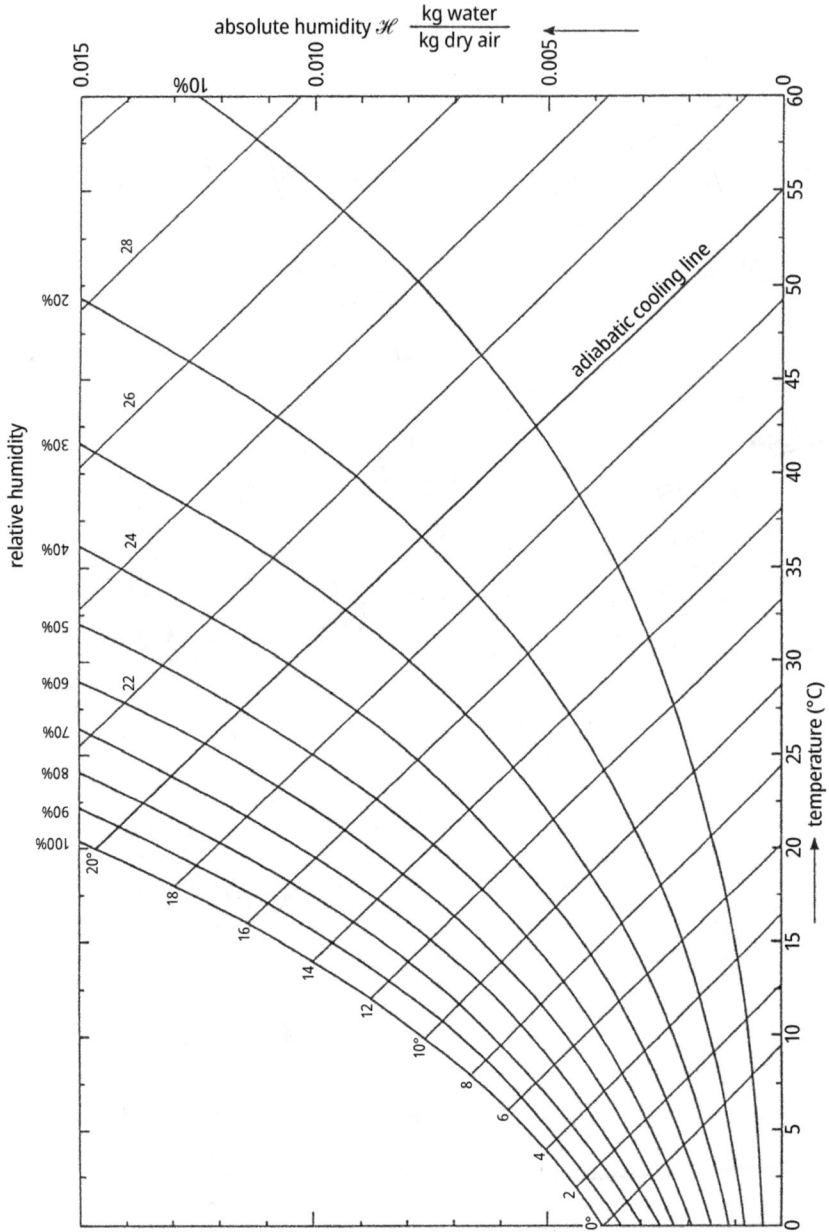

Fig. 4.29: Humidity diagram for air water systems (ambient temperatures).

Fourier instationary heat and mass transfer

Heat transfer

Center temperature (T_c) as a function of time (t):

$$Mc = \frac{T_1 - T_c}{T_1 - T_0} \quad \text{(fig pp 116)} \quad \text{and} \quad Fo = \frac{at}{d^2}$$

Average temperature $(<T>)$ as a function of time (t):

$$\langle M \rangle = \frac{T_1 - \langle T \rangle}{T_1 - T_0} \quad \text{(fig pp 117)} \quad \text{and} \quad Fo = \frac{at}{d^2}$$

T_1 is the boundary temperature, T_0 is the initial temperature (at $t = 0$), T_c is the temperature at the center, $\langle T \rangle$ is the average body temperature, a is the thermal diffusivity (m^2/s), and d is the characteristic dimension of the object (m).

Mass transfer

Center concentration (C_c) as a function of time (t):

$$Mc = \frac{C_1 - C_c}{C_1 - C_0} \quad \text{(fig pp 116)} \quad \text{and} \quad Fo = \frac{\mathcal{D}t}{d^2}$$

Average concentration $(<C>)$ as a function of time (t):

$$\langle M \rangle = \frac{C_1 - \langle C \rangle}{C_1 - C_0} \quad \text{(fig pp 117)} \quad \text{and} \quad Fo = \frac{\mathcal{D}t}{d^2}$$

where C_1 is the boundary concentration, C_0 is the initial concentration (at $t = 0$), C_c is the concentration at the center, $\langle C \rangle$ is the average concentration, $\mathcal{D}$ is the diffusion coefficient (m^2/s), and d is the characteristic dimension of the object (m).

Fourier instationary heat and mass transfer (center *T* or *C*)

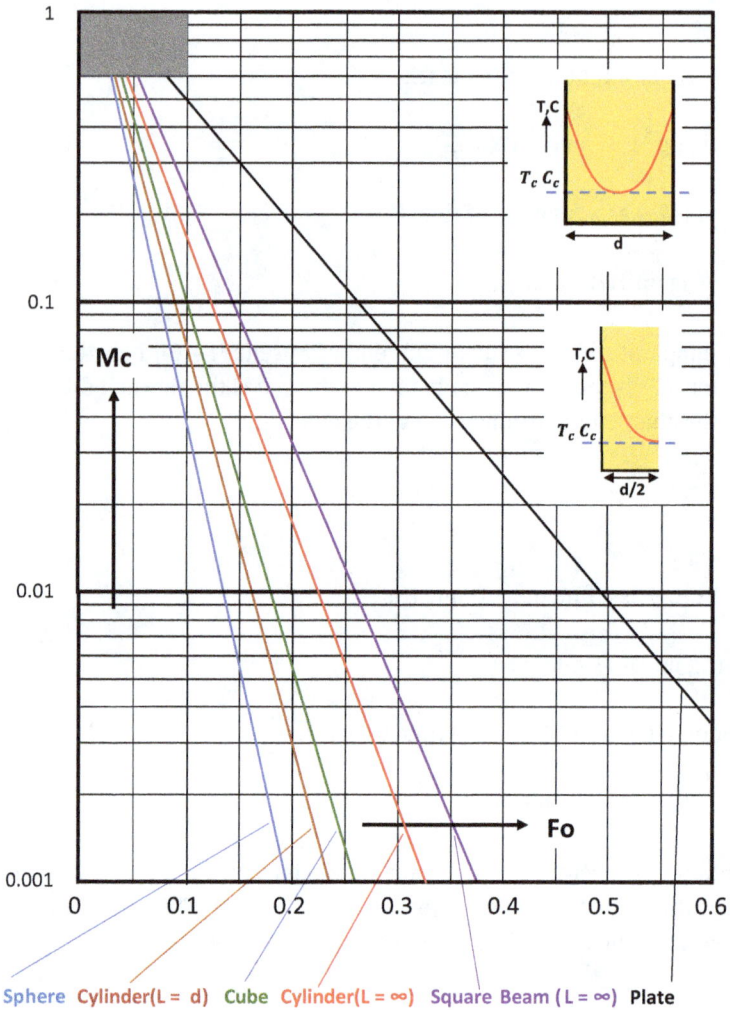

Fig. 4.30: Center temperature or concentration for instationary processes.

Fourier instationary heat and mass transfer (average *T* or *C*)

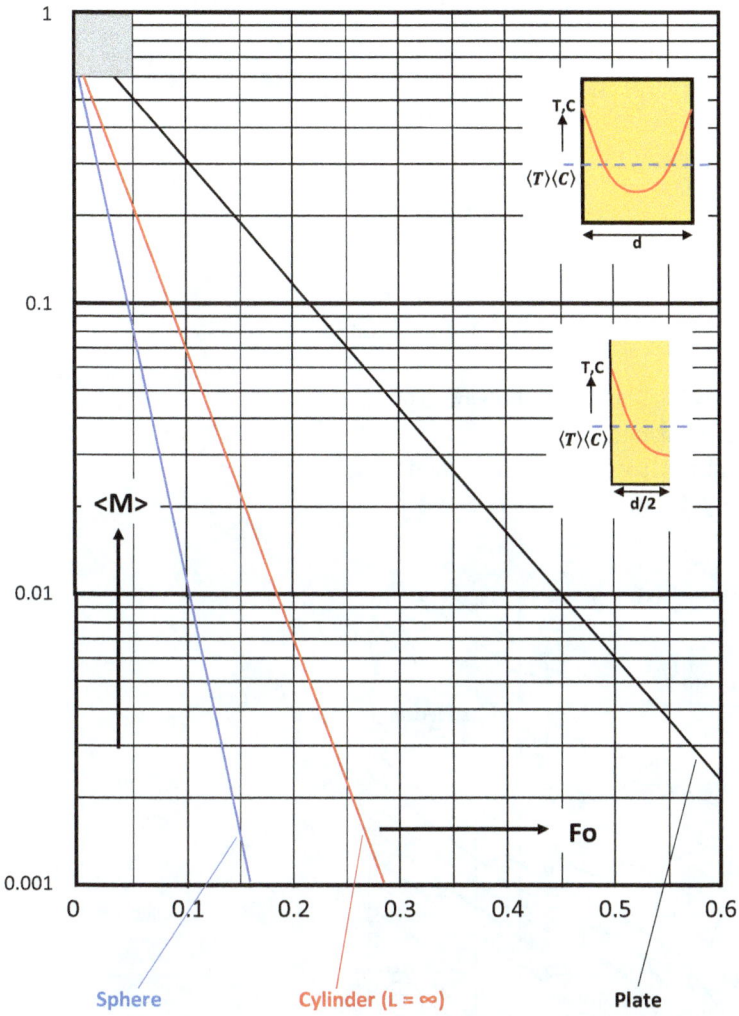

Fig. 4.31: Average temperature or concentration for instationary processes.

Shapes of free rising bubbles or drops in Newtonian liquids

$$\mathrm{Re} = \frac{\rho_L d_e v}{\eta_L}$$

$$\mathrm{Eo} = \frac{g d_e^2}{\sigma} \Delta\rho$$

$$\mathrm{Mo} = \frac{g \eta_L^4}{\rho_L^2 \sigma^3} \Delta\rho$$

where

ρ_L is the density of the fluid,

d_e is the volume equivalent diameter of the bubble or drop,

v is the rising velocity of the bubble or drop,

σ is the surface tension,

η_L is the dynamic viscosity of the fluid, and

$\Delta\rho$ is the density difference between phases.

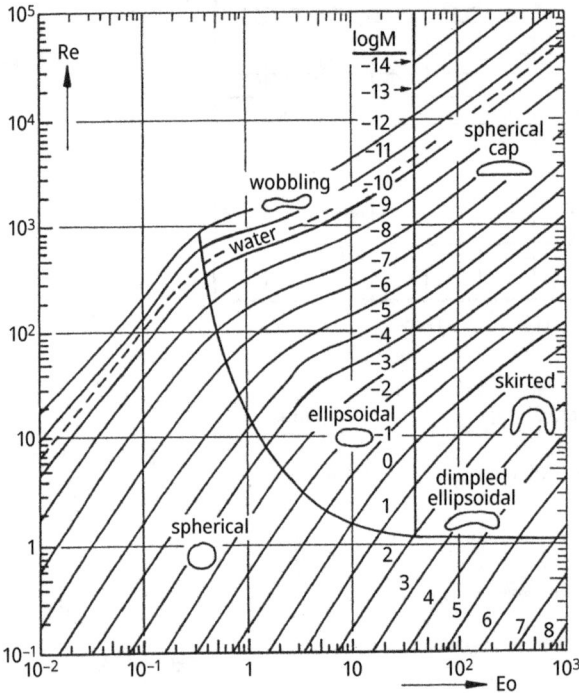

Fig. 4.32: Rising bubbles and drops: Reynolds number versus Eötvös number at different Morton numbers.

Rising Velocity of Air Bubbles in Water

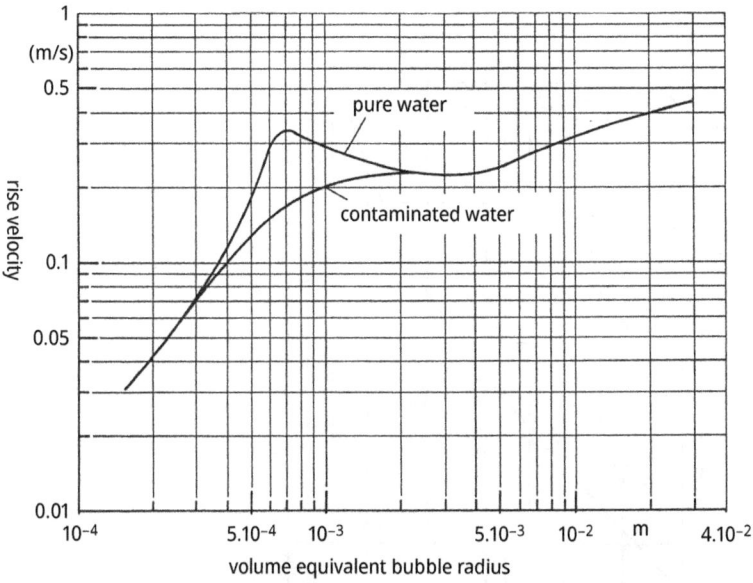

Fig. 4.33: Influence of water quality on rising air bubbles.

Rising velocity or fall velocity of drops in liquids of low viscosity $\eta_c < 5$ mPa s

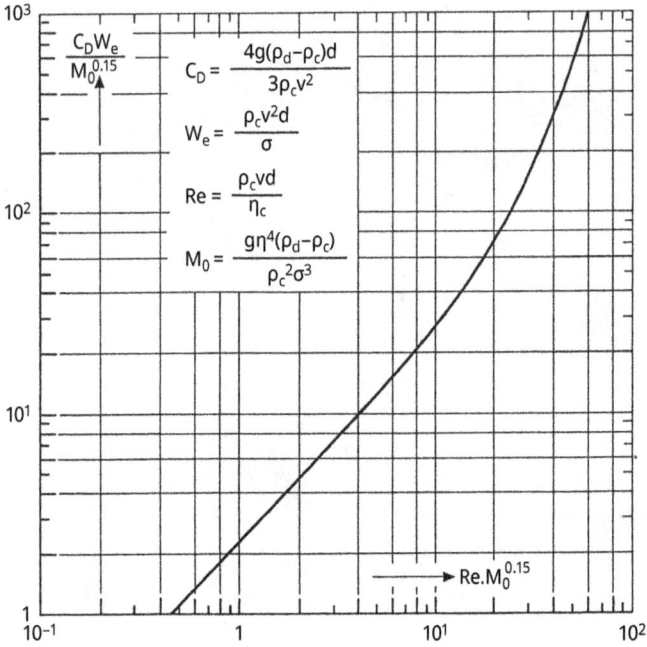

Inside the figure:

$$\frac{C_D W_e}{M_0^{0.15}}$$

$$C_D = \frac{4g(\rho_d - \rho_c)d}{3\rho_c v^2}$$

$$W_e = \frac{\rho_c v^2 d}{\sigma}$$

$$Re = \frac{\rho_c v d}{\eta_c}$$

$$M_0 = \frac{g\eta^4(\rho_d - \rho_c)}{\rho_c^2 \sigma^3}$$

$$Re.M_0^{0.15}$$

Fig. 4.34: Velocity of drops in liquids of low viscosity.

Two-phase gas-liquid flow in horizontal tubes

Parameter	Definition	Unit
Φ_g	= mass flux of the gas	kg/m^2 s
Φ_L	= mass flux of the liquid	kg/m^2 s
λ	$= 0.029\sqrt{\rho_g \rho_L}$	
ψ	$= \frac{73}{\sigma}\sqrt[3]{\frac{\eta_L}{\rho_L}}$	
ρ_g	= density of the gas	kg/m^3
ρ_L	= density of the fluid	kg/m^3
η_L	= viscosity of the fluid	Pa s
σ	= surface tension between gas and fluid	N/m
1	Stratified	
2	Wave	
3	Annular	
4	Spray or dispersed drops	
5	Froth or dispersed bubbles	
6	Slugs	
7	Plugs	

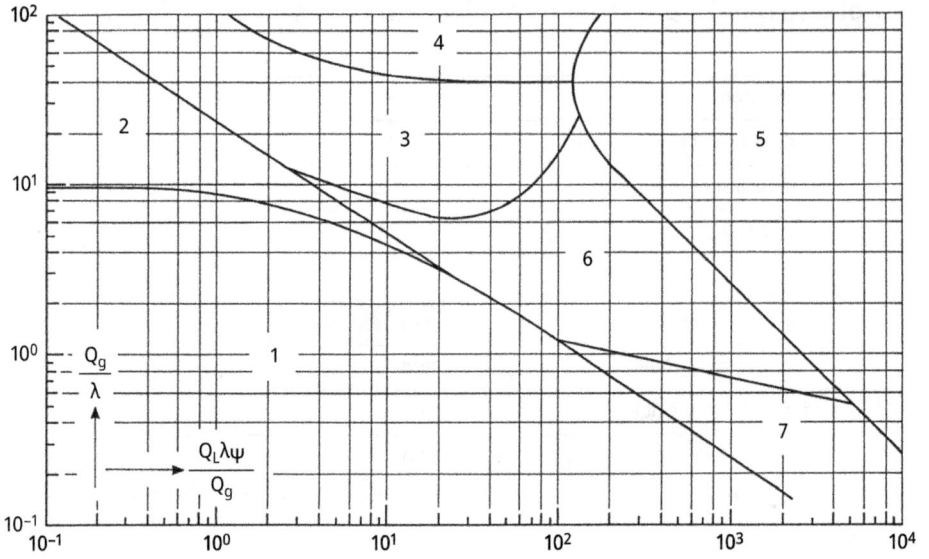

Fig. 4.35: Two phase flow in horizontal pipes.

Two-phase cocurrent flow of gas-liquid through vertical tubes

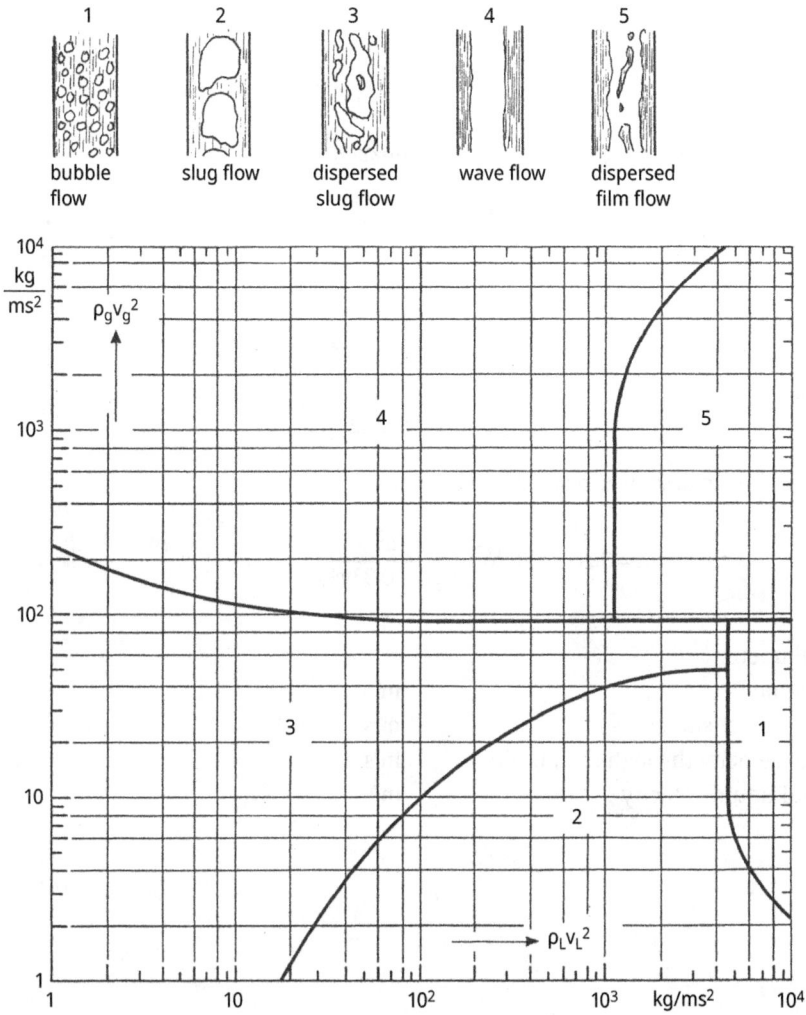

1 bubble flow
2 slug flow
3 dispersed slug flow
4 wave flow
5 dispersed film flow

Fig. 4.36: Two phase flow in vertical pipes.

Formation of bubbles at a nozzle

In stationary liquid

Single bubbles:

$$d = 5.55 \times 10^{-3} \left(u.\delta^{2.5}\right)^{0.289}$$

Coalescing bubbles (coalescing directly after the nozzle):

$$d = 6.30 \times 10^{-3} \left(u.\delta^{2.5}\right)^{0.305}$$

In flowing liquid

Single bubbles:

$$d = 1.55 \times 10^{-2} \cdot \delta \left(\frac{u^2}{gd + 0.33v^2}\right)^{0.2}$$

Coalescing bubbles (coalescing directly after the nozzle):

$$d = 3.20 \times 10^{-2}\delta^{1.1} \left(\frac{u^2}{gd + 0.33v^2}\right)^{0.2}$$

d = bubble diameter   m
δ = nozzle diameter   m
g = gravitational constant   m/s^2
u = linear gas velocity through the nozzle   m/s
v = superficial liquid velocity   m/s

Mass transfer with first-order chemical reaction

$$Ha = \frac{\sqrt{k_r\,\mathcal{D}_A}}{k} \ll 1 \quad \frac{A}{V}\frac{k}{k_r} \gg 1$$

No reaction in the boundary layer

$$\Phi''_A = \frac{V}{A}k_r\,c_{Ai} \quad c_A \approx c_{Ai}$$

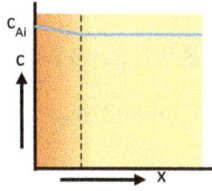

$$Ha = \frac{\sqrt{k_r\,\mathcal{D}_A}}{k} \ll 1 \quad \frac{A}{V}\frac{k}{k_r} \approx 1$$

No reaction in the boundary layer

$$\Phi''_A = \frac{V}{A}k_r c_{Ai}\left\{\frac{Ak}{Ak + Vk_r}\right\} = k\,c_{Ai}\left\{\frac{Vk_r}{Vk_r + Ak}\right\}$$

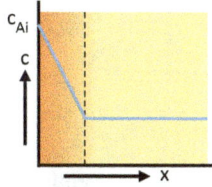

$$Ha = \frac{\sqrt{k_r\,\mathcal{D}_A}}{k} \ll 1 \quad \frac{A}{V}\frac{k}{k_r} \ll 1$$

No reaction in the boundary layer

$$\Phi''_A = k\,c_{Ai} \quad c_A \approx 0$$

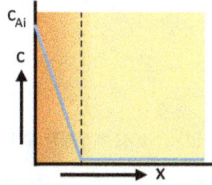

$$Ha = \frac{\sqrt{k_r\,\mathcal{D}_A}}{k} \approx 1 \quad \frac{A}{V}\frac{k}{k_r} \ll 1$$

Partial reaction in the boundary layer

$$\Phi''_A = \sqrt{k_r\,\mathcal{D}_A + k^2}\cdot c_{Ai} \quad c_A \approx 0$$

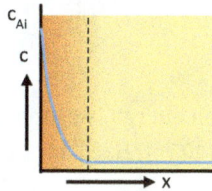

$$Ha = \frac{\sqrt{k_r\,\mathcal{D}_A}}{k} \gg 1 \quad \frac{A}{V}\frac{k}{k_r} \ll 1$$

Reaction completely in the boundary layer

$$\Phi''_A = \sqrt{k_r\,\mathcal{D}_A}\cdot c_{Ai} \quad c_A = 0$$

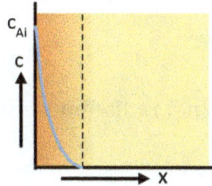

Symbols

Ha	Hatta number	V	Volume
k_r	Reaction constant	A	Interfacial area
k	Mass transfer coefficient	c_A	Concentration of A
$\mathcal{D}_A$	Diffusion coefficient of A	c_{Ai}	Interfacial concentration of A
Φ''	Flux	x	Distance from the wall

Ref: Fysische transportverschijnselen I, Smith,J.M., Janssen,L.P.B.M. and Stammers,E.J., Delftse Uitgevers Maatschappij (1981)

Radiation

Stefan-Boltzmann law

$$\Phi_r{}'' = \varepsilon \cdot \sigma T^4 \, (T \text{ in kelvin})$$

$\sigma = 5.670374 \times 10^{-8} \, \text{W/m}^2 \, \text{K}^4$ (Stefan-Boltzmann constant);
$0 \le \varepsilon \le 1$ (emissivity)

Wien's displacement law

$$\lambda_{\max} = \frac{b}{T} \quad (T \text{ in kelvin})$$

$b = 2.897772 \times 10^{-3} \text{m} \cdot \text{K} \;$ (Wien's displacement constant)

Kirchhoff's law

$$\Phi_n{}'' = \varepsilon \; \Phi_b{}'' - a \, \Phi_i{}'' \;\; (\text{net radiation} = \text{own emission} - \text{absorbed radiation})$$

Planck's law

$$E = hf \; (\text{photon energy is proportional to frequency})$$
$$h = 6.62607 \times 10^{-34} \text{J/Hz} (\text{Planck's constant})$$

Blackbody radiation exchange

$$\Phi_b{}'' = \sigma \left(T_1^4 - T_2^4 \right) (T \text{ in kelvin}) \; (\text{net radiation is radiation out} - \text{radiation in})$$

Wavelength-temperature scale of light sources

Light sources	Wavelength (μm)	Color temperature (K)
Flame of a match stick	1.70	1,700
Candle flame, sunset/sunrise	1.61	1,800
Incandescent light bulb	0.88–1.07	2,700–3,300
Studio lamps, photofloods, etc.	0.91	3,200
Carbon arc, acetylene oxygen flame	0.78	3,700
Moonlight, xenon arc lamp	0.71	4,100
Horizon daylight	0.58	5,000
Vertical daylight, electronic flash	0.48–0.53	5,500–6,000
Overcast daylight	0.45	6,500
LCD or CRT screen	0.31–0.45	6,500–9,300
Residential lightening		
Warm white	1.07	2,700
Neutral white	0.83–0.97	3,000–3,500
Cool white	0.71	4,100
Daylight	0.45–0.58	5,000–6,500

Color temperatures from: Asim Kumar Roy Choudhury: Principles of Colour and appearance Measurement (2014).

Color temperature scale (in kelvin)

1000K 2000K 3000K 4000K 5000K 6000K 7000K 8000K 9000K 10000K

Fig. 4.37: Colour as function of temperature.

The electromagnetic spectrum

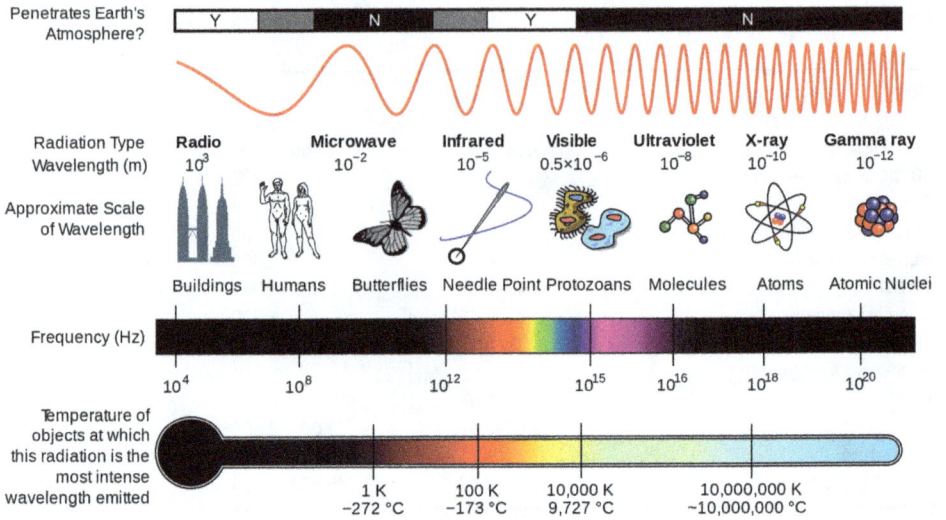

Fig. 4.38: Electromagnetic wave lengths.
ref: NASA/Wikipedia: CC BY_SA 3.0; file:EM spectrum3-new.jpg

Emission coefficients of various materials

Based on a temperature of 300 K

Aluminum paint	0.3–0.4
Aluminum polished	0.053
Aluminum sheet	0.071
Asbestos	0.96
Black colored bodies	1.00
Black lacquer	0.83
Brass polished	0.050
Brass sheet	0.069
Coal	0.81
Copper polished	0.041
Enamel (white)	0.90
Glass	0.94
Gypsum	0.90
Ice	0.91
Iron (cast)	0.81
Iron (nickel coated)	0.059
Iron plate, rusted	0.69
Marble polished	0.85

(continued)

Paint (enamel)	0.91
Paint (lacquer)	0.90
Paints (oil)	0.89–0.97
Precious metals polished	0.02–0.05
Red brick	0.94–0.95
Roofing paper	0.91
Rubber (hard)	0.95
Rubber (soft)	0.87
Steel (ground sheet)	0.66
Tin-coated iron	0.057
Water	0.91
Wood (oak)	0.90

List of symbols and abbreviations

a	Absorption coefficient	–
b	Wien's displacement constant	$m \cdot K$
E	Radiation energy	J
f	Frequency	Hz
h	Planck's constant	J/Hz
T	Absolute temperature	K
ε	Emission coefficient	–
λ_{max}	Wavelength at the maximum	m
σ	Stefan-Boltzmann constant	$W/m^2\ K^4$
Φ_b''	Energy flux of blackbody radiation	W/m^2
Φ_i''	Energy flux of blackbody absorption	W/m^2
Φ_n''	Net energy flux by radiation	W/m^2

Environmental data

General

Earth

Volume of the oceans	1.35×10^9 km^3	97% of all water
Surface of the oceans	361×10^6 km^2	71% of the Earth's surface
Average depth of the oceans	3,700 m	
Depth of thermocline	500 m	(mixing barrier)
Density of seawater	1,023.6 kg/m^3	at 1 bar, 25 °C,
Salinity of seawater	35 kg/m^3	= 35,000 ppm
Mass of the Earth	5.972×10^{24} kg	
Average distance to the Sun	152×10^6 km	
Forrest land/total land area	0.307	
Agricultural land/total land area	0.374	

Atmosphere

Total mass of the atmosphere	5×10^{18} kg	
Mass of the atmosphere below 11 km	3.75×10^{18} kg	= 75% of total atmospheric mass
CO_2 content in the atmosphere	400 ppm	
CO_2 emission into the atmosphere	760×10^{12} kg/year	(2019)

Sun

Mass of the Sun	1.9885×10^{30} kg	mass ratio of the Sun/Earth = 333,000
Temperature of the Sun	5,500 °C	Temperature photosphere
Energy emission of the Sun	62.94×10^6 J/m^2·s	
Irradiation of sunlight:		
At the top of atmosphere	1,361 W/m^2	
At the Earth's surface (max)	1,120 W/m^2	Perpendicular surface
Daily average at ground	21.6 MJ/m^2	= 6 kWh/m^2

Atmosphere

CO_2 data for different fuels

Adapted from: Emission factors for Greenhouse Gas Inventories Environmental Protection Agency, 2014.

	CO_2 emission (kg$_{CO2}$/kg$_{fuel}$)	Combustion energy (MJ/kg$_{fuel}$)	Specific CO_2 emission (kg$_{CO2}$/GJ)
Hydrogen	0	141.58	0
Natural gas (methane)	2.75	55.51	50
Methanol	1.37	22.65	60
Bioalcohol (ethanol)	1.91	29.67	64
Kerosene	3.00	44.53	67
Diesel	3.15	45.72	69
Gasoline (octane)	3.30	47.87	71
Coal (anthracite)	3.37	32.40	104
Coal (coke)	2.82	25.92	109
Wood	1.83	16.20	113
Peat	1.91	13.05	146

Greenhouse effect of some gasses (1 kg gas is equivalent to X kg CO_2)

Gas		X kg CO_2
Carbon dioxide	CO_2	1
Methane	CH_4	25
Nitrous oxide	N_2O	298
Difluoromethane	CH_2F_2	675
Tetrafluoromethane	CF_4	7,390
Hexafluoroethane	C_2F_6	12,200
Octafluoropropane	C_3F_8	8,830
Perfluorobutane	C_4F_{10}	8,860
Sulfur hexafluoride	SF_6	22,800

Solar energy

Insolation in peak sun hours per year (1 PSH ≡ 1 kWh/m^2)

Continent	Location	Peak sun hours/year
Europe	North	1,000–1,200
	Middle	1,200–1,400
	South	1,400–1,700
North America	Canada	1,100–1,400
	USA North	1,400–1,700
	USA South	1,700–2,100
	Mexico	1,800–2,200
Central America	–	1,700–2,100
South America	North	1,500–2,000
	Middle, East	1,500–2,000
	Middle, West	2,400–2,600
	South	1,100–1,600
Australia	North and Middle	2,200–2,400
	South (Inlands)	1,900–2,200
	South (Coast)	1,500–1,600
Asia	Southwest	2,000–2,300
	India and Himalayas	1,800–2,100
	China North and Mongolia	1,500–1,700
	China South	1,100–1,300
	Southeast Asia	1,600–1,800
	Indonesia and Philippines	1,600–1,900
Africa	North	2,400–2,700
	Middle	1,600–2,200
	South	1,300–2,400

Derived from Global Solar Atlas.

Conversion factors of solar energy *

	Peak sun hours/day	Peak sun hours/year	W/m^2	kWh/m^2 day	kWh/m^2 year
Peak sun hours/day	1	365.24	41.667	1	365.24
Peak sun hours/year	0.002738	1	0.1141	0.002738	1
W/m^2	0.0240	8.7658	1	0.0240	8.7658
kWh/m^2 day	1	365.24	41.667	1	365.24
kWh/m^2 year	0.002738	1	0.1141	0.002738	1

*Key: 1 PSH/day = 365.24 PSH/year = 41.667 W/m^2 = 1 kWh/m^2 day = 365.24 kWh/m^2 year.

Energy storage

			Typical energy densities	
			(kJ/kg)	(MJ/m³)
Thermal energy	Water, temperature difference ΔT: 100–40 °C		252	246
	Water/steam phase change at constant temperature	100 °C 140 °C 200 °C 300 °C	2,258 2,149 1,942 1,404	–
	Stone/rocks (granite–dolomite) ΔT: 100–40 °C ΔT: 400–200 °C		46 152	126 420
	Clay (dry) ΔT: 100–40 °C ΔT: 400–200 °C		59 197	95 317
	Sand (packed) ΔT: 100–40 °C ΔT: 400–200 °C		50 160	80 250
	Iron ΔT: 100–40 °C ΔT: 400–200 °C		30 100	230 800
Combustion energy	Crude oil		42,000	37,000
	Coal		32,000	42,000
	Dry wood		12,500–20,000	10,000–16,000
	Hydrogen, gas (LHV/HHV)		120,000/142,000	11/13
	Liquified petroleum gas (LPG) (LHV/HHV)		46,000/51,000	21,000/24,000
	Methanol		21,000	17,000
	Natural gas (LHV/HHV)		42,000/55,000	30/37
	Ethanol		28,000	22,000
Electrochemical energy	Lead-acid batteries		200	365
	Nickel-cadmium batteries		750	600
	Lithium ion batteries		940	777

(continued)

		Typical energy densities	
		(kJ/kg)	(MJ/m³)
Mechanical energy	Hydropower, 100 m head	0.98	0.98
	Compressed air at 60 bar ato	–	23

Solid State Ionics 134 (2000) 139, Dell, R.M.
Energy Environ. Sci. (2011), 4, 2614, Z. Chen-Xi, L. Hong.

Heating values

$$\text{HHV} = \text{LHV} + H_{v,\,H_2O} \times \left(\frac{n_{H_2O,out}}{n_{fuel,in}} \right)$$

where HHV is the higher heating value, LHV is the lower heating value, H_v is the heat of vaporization, and n is the number of moles.

Chapter 5
Material properties

Water

- Prandtl
- Thermal diffusion coefficient
- Heat of vaporization
- Cubic expansion coefficient
- Dielectrical constant
- Surface tension
- Electrical conductivity
- Density
- Vapor pressure
- Thermal conductivity coefficient
- Dynamic viscosity
- Specific heat
- Compressibility
- Refractive index
- Velocity of sound
- Properties of water and steam above 100 °C

https://doi.org/10.1515/9783111385341-005

The Prandtl number $Pr = \frac{\eta c_p}{\lambda}$, for water (up to 100 °C at atmospheric pressure, and above 100 °C at vapor pressure)

Thermal diffusion coefficient, $a = \frac{\lambda}{\rho c_p}$, for water in m^2/s (up to 100°C at atmospheric pressure, and above 100 °C at vapor pressure)

T (°C)	Pr	T (°C)	a
0	13.5	0	0.131×10^{-6}
10	9.45	10	0.138×10^{-6}
20	7.01	20	0.143×10^{-6}
30	5.45	30	0.147×10^{-6}
40	4.34	40	0.151×10^{-6}
50	3.58	50	$0.155 \cdot \times 10^{-6}$
60	3.00	60	0.158×10^{-6}
70	2.57	70	0.161×10^{-6}
80	2.23	80	0.164×10^{-6}
90	1.96	90	0.166×10^{-6}
100	1.75	100	0.168×10^{-6}
120	1.45	120	0.171×10^{-6}
140	1.25	140	0.172×10^{-6}
160	1.09	160	0.174×10^{-6}
180	0.98	180	0.173×10^{-6}
200	0.92	200	0.171×10^{-6}
220	0.88	220	0.168×10^{-6}
240	0.87	240	0.164×10^{-6}
260	0.87	260	0.156×10^{-6}
280	0.91	280	0.145×10^{-6}
300	1.00	300	0.129×10^{-6}
320	1.15	320	0.111×10^{-6}
340	1.51	340	0.084×10^{-6}
360	2.62	360	0.050×10^{-6}

Heat of vaporization, ι, of water in kJ/kg, at vapor pressure.

Cubic expansion coefficient, $\gamma = \frac{1}{V}\left(\frac{\partial V}{\partial T}\right)$ of water in K^{-1} (up to 100 °C at atmospheric pressure, and above 100 °C at vapor pressure).

T (°C)	ι	T (°C)	γ
0	2,502.2	0	-0.07×10^{-3}
10	2,478.6	10	0.088×10^{-3}
20	2,455.2	20	0.206×10^{-3}
30	2,431.9	30	0.303×10^{-3}
40	2,408.1	40	0.385×10^{-3}
50	2,384.3	50	0.457×10^{-3}
60	2,360.0	60	0.523×10^{-3}
70	2,335.4	70	0.585×10^{-3}
80	2,310.1	80	0.643×10^{-3}
90	2,284.1	90	0.698×10^{-3}
100	2,257.8	100	0.753×10^{-3}
120	2,202.6	120	0.860×10^{-3}
140	2,144.9	140	0.975×10^{-3}
160	2,082.2	160	1.098×10^{-3}
180	2,013.4	180	1.233×10^{-3}
200	1,942.0	200	1.392×10^{-3}
220	1,859.9	220	1.597×10^{-3}
240	1,767.7	240	1.862×10^{-3}
260	1,662.6	260	2.21×10^{-3}
280	1,552.7	280	2.70×10^{-3}
300	1,404.1	300	3.46×10^{-3}
320	1,236.9	320	4.60×10^{-3}
340	1,026.2	340	8.25×10^{-3}
360	720.2		

Dielectrical constant ϵ of water (up to 100 °C at atmospheric pressure, and above 100 °C at vapor pressure).

T (°C)	ϵ
0	87.69
10	83.82
20	80.08
30	76.49
40	73.02
50	69.07
60	66.51
70	63.45
80	60.54
90	57.77
100	55.15
120	50.48
140	46.00
160	41.87
180	38.10
200	34,59
220	31.32
240	28.24
260	25.29
280	22.45
300	19.66
340	14.10
370	9.74

Surface tension τ of water against air in N/m.

T (°C)	σ
0	75.64×10^{-3}
5	74.9×10^{-3}
10	74.22×10^{-3}
15	73.49×10^{-3}
18	73.05×10^{-3}
20	72.75×10^{-3}
25	71.79×10^{-3}
30	71.18×10^{-3}
40	69.56×10^{-3}
50	$67,91 \times 10^{-3}$
60	$66.18 \cdot \times 10^{-3}$
70	64.42×10^{-3}
80	62.61×10^{-3}
90	60.75×10^{-3}
100	58.85×10^{-3}

Electrical conductivity γ of water in S/m $(= \Omega^{-1}\, m^{-1})$.

T (°C)	γ
−2	1.47×10^{-6}
0	1.58×10^{-6}
2	1.80×10^{-6}
4	2.12×10^{-6}
10	2.85×10^{-6}
18	4.41×10^{-6}
26	6.70×10^{-6}
34	9.62×10^{-6}
50	18.9×10^{-6}

Density ρ of water in kg/m³ at atmospheric pressure.

Temperature (°C)	0	1	2	3	4	5	6	7	8	9
0	999.87	999.93	999.97	999.99	1,000.00	999.99	999.97	999.93	999.88	999.81
10	999.73	999.63	999.52	999.40	999.27	999.13	998.97	998.80	998.62	998.43
20	998.23	998.02	997.80	997.57	997.33	997.08	996.82	996.55	996.55	995.98
30	995.68	995.37	995.06	994.73	994.40	994.06	993.71	993.36	993.00	992.63
40	992.25	991.87	991.47	991.07	990.66	990.25	989.82	989.40	988.96	988.52
50	988.07	987.61	987.15	986.68	986.21	985.72	985.24	984.74	984.24	983.74
60	983.23	982.71	982.19	981.66	981.12	980.58	980.03	979.48	978.92	978.36
70	977.93	977.22	976.64	976.06	975.47	974.87	974.27	973.67	973.06	972.44
80	971.82	971.19	970.56	969.93	969.29	968.64	967.00	967.34	966.74	966.01
90	965.34	964.67	963.99	963.30	962.61	961.92	961.22	960.52	959.81	959.10
100	958.38									

Vapor pressure p_{w} of water in Pa.

Temperature (°C)	0	1	2	3	4	5	6	7	8	9
0	610.3	656.6	705.6	757.8	813.2	872.1	934.8	1,001.4	1,072.3	1,147.5
10	1,227.5	1,312.1	1,401.9	1,497.0	1,597.7	1,704.5	1,817.3	1,936.7	2,062.9	2,196.2
20	2,337.2	2,485.8	2,642.7	2,808.1	2,982.6	3,166.4	3,360.1	3,564.0	3,778.6	4,004.4
30	4,241.8	4,491.2	4,753.5	5,028.9	5,318.0	5,621.5	5,939.8	6,273.5	6,615.0	6,990.0
40	7,374.1	7,776	8,197	8,637	9,098	9,581	10,083	10,610	11,158	11,732
50	12,331	12,956	13,608	14,289	14,996	15,733	16,501	17,304	18,138	19,007
60	19,911	20,850	21,829	22,843	23,900	24,997	26,137	27,319	28,547	29,821
70	31,150	32,510	33,940	35,420	36,950	38,530	40,170	41,870	43,630	45,450
80	47,330	49,280	51,300	53,400	55,560	57,790	60,100	62,470	64,930	67,460
90	70,078	72,783	75,574	78,454	81,427	84,492	87,654	90,913	94,272	97,733
100	101,300									

Thermal conductivity coefficient λ of water in W/mK.

T (°C) / P (Pa)	0	25	50	75	100	150	200	250	300
10^5	0.552	0.607	0.641	0.665	0.682				
10^7	0.555	0.609	0.644	0.669	0.688	0.692	0.672	0.628	0.543
2×10^7	0.557	0.613	0.648	0.673	0.695	0.699	0.680	0.636	0.557
3×10^7		0.616	0.652	0.677	0.702	0.706	0.688	0.648	0.576
4×10^4		0.620	0.656	0.682	0.709	0.714	0.698	0.658	0.595

Dynamic viscosity η of water in mPa s.

P (Pa) T (°C)	10^5	5×10^6	10^7	2×10^7	3×10^7
0	1.792	1.781	1.770	1.748	1.726
10	1.307	1.301	1.296	1.289	1.281
20	1.002	1.001	1.000	0.998	0.995
30	0.797	0.797	0.789	0.798	0.800
40	0.653	0.653	0.654	0.656	0.658
50	0.546	0.547	0.549	0.552	0.555
60	0.466	0.468	0.469	0.472	0.467
70	0.404	0.406	0.408	0.411	0.416
80	0.355	0.358	0.361	0.366	0.372
90	0.315	0.319	0.324	0.330	0.337
100	0.282	0.287	0.293	0.301	0.309
120		0.228	0.245	0.253	0.262
140		0.202	0.207	0.216	0.224
160		0.175	0.178	0.185	0.193
180		0.154	0.157	0.163	0.169
200		0.139	0.141	0.145	0.149
220		0.127	0.129	0.131	0.134
240		0.116	0.118	0.120	0.123
260		0.108	0.109	0.111	0.114
280			0.101	0.103	0.106
300			0.094	0.096	0.099
320				0.089	0.093
340				0.079	0.085
360				0.067	0.077
380					0.065
400					0.043

Specific heat C_p of water in kJ/kg K.

P (Pa) T (°C)	10^5	5×10^6	10^7	1.5×10^7	2×10^7	2.5×10^7	3×10^7
0	4.2210	4.207	4.198	4.190	4.182	4.173	4.165
20	4.1850	4.173	4.165	4.156	4.144	4.136	4.123
40	4.1816	4.165	4.156	4.144	4.131	4.123	4.110
60	4.1875	4.169	4.156	4.144	4.131	4.119	4.106
80	4.1996	4.186	4.169	4.156	4.144	4.127	4.115
100	4.2193	4.207	4.190	4.177	4.161	4.144	4.131
120		4.236	4.219	4.203	4.186	4.169	4.152
140		4.270	4.253	4.228	4.215	4.198	4.177
160		4.328	4.307	4.286	4.265	4.244	4.224
180		4.400	4.374	4.349	4.324	4.303	4.278
200		4.487	4.458	4.429	4.400	4.370	4.345

(continued)

P (Pa) 10^5 / T (°C)	5×10^6	10^7	1.5×10^7	2×10^7	2.5×10^7	3×10^7
220	4.596	4.559	4.525	4.492	4.458	4.425
240	4.743	4.697	4.651	4.609	4.567	4.529
260	4.948	4.886	4.827	4.772	4.722	4.668
280		5.158	5.078	5.003	4.932	4.865
300		5.552	5.447	5.305	5.204	5.124
320			6.201	5.828	5.677	5.439
340			8.124	7.018	6.407	5.971

Compressibility $k = -\frac{1}{V}\left(\frac{\partial V}{\partial P}\right)$ of water in 10^{-12} Pa^{-1}.

T (°C) / P (Pa)	0	20	0	100	150	200	250	300	350
10^2	506	434	450	480					
5×10^3	497	425	439	474	592	832	1,394		
10^4	488	419	428	466	576	806	1,327	3,000	
1.5×10^4	480	415	418	460	562	781	1,260	2,650	
2×10^4	473	412	410	452	550	758	1,193	2,350	9,200
2.5×10^4	465	410	403	445	540	735	1,130	2,120	7,100
3×10^4	458	407	397	439	531	713	1,072	1,920	5,330
3.5×10^4	452	404	393	420	523	693	1,020	1,740	
4.5×10^4	440	398	384	385	511	654	925	1,420	
7.5×10^4	406	383	359	360					
10^5	378	373	340						

Refractive index η of water.

T (°C) / λ (µm)	15	20	25	30	40
0.70652	1.331409	1.330019	1.329544	1.328993	1.327685
0.66871	1.331269	1.330876	1.330398	1.329843	1.328528
0.65628	1.331545	1.331151	1.330672	1.330116	1.328798
0.58926	1.333387	1.332988	1.332503	1.331940	1.330610
0.57696	1.333781	1.33338	1.332894	1.332331	1.330998
0.54607	1.334869	1.334466	1.333977	1.333411	1.332071
0.50157	1.336760	1.336353	1.335860	1.335289	1.33936
0.48613	1.337531	1.337123	1.336628	1.336055	1.334702
0.47131	1.338341	1.337931	1.336860	1.336860	1,335504
0.44715	1.339835	1.339423	1.338925	1.338347	1.336984
0.43583	1.340626	1.340210	1.339716	1.339131	1.337765
0.40466	1.34158	1.342742	1.342239	1.341656	1.340280

Velocity of sound c in water in m/s.

T (°C)	c (m/s)
0	1,402.7
5	1,426.5
10	1,447.6
15	1,466.3
20	1,482.7
25	1,497.0
30	1,509.4
40	1,529.2
50	1,542.9
60	1,551.3
80	1,554.8
100	1,543.4

Properties of water and steam above 100 °C.

Temperature (°C)	Pressure (kPa)	Density Liquid (kg/m³)	Density Vapor (kg/m³)	Enthalpy Liquid (kJ/kg)	Enthalpy Vapor (kJ/kg)	Heat of vaporization (kJ/kg)	Entropy Liquid (kJ/kg °C)	Entropy Vapor (kJ/kg °C)	Vaporization entropy (kJ/kg °C)
100	101.3	958.3	0.598	419.17	2,677.0	2,257.8	1.3065	7.3510	6.0445
110	143.3	951.0	0.827	461.40	2,692.1	2,230.7	1.4178	7.2342	5.8164
120	198.5	943.1	1.122	504.06	2,706.7	2,202.6	1.5266	7.1237	5.5971
130	270.1	934.8	1.496	546.38	2,720.6	2,174.2	1.6330	7.0208	5.3878
140	361.4	926.1	1.967	589.11	2,734.0	2,144.9	1.7372	6.9233	5.1860
150	476	916.9	2.548	632.27	2,746.5	2,114.2	1.8397	6.8315	4.9918
160	618	907.4	3.260	675.85	2,758.3	2,082.2	1.9410	6.7445	4.8035
170	792	897.3	4.122	719.42	2,769.2	2,049.8	2.0403	6.6616	4.6213
180	1,003	886.9	5.157	763.42	2,778.8	2,013.4	2.1378	6.5808	4.4430
190	1,255	876.0	6.392	807.83	2,787.6	1,978.8	2.2336	6.5042	4.2706
200	1,555	864.7	7.857	852.66	2,794.7	1,942.0	2.3283	6.4289	4.1006
210	1,908	864.7	9.585	897.92	2,800.2	1,902.3	2.4229	6.3560	3.9331
220	2,320	840.3	11.61	944.01	2,803.9	1,859.9	2.5158	6.2840	3.7682
230	2,798	827.3	13.98	1,020.52	2,806.0	1,786.0	2.6075	6.2120	3.6045
240	3,348	813.6	16.75	1,037.86	2,805.6	1,767.7	2.6991	6.1404	3.4413
250	3,978	799.2	19.98	1,086.05	2,803.1	1,717.1	2.7908	6.0693	3.2785
260	4,694	784.0	23.74	1,135.49	2,798.1	1,662.6	2.8825	5.9973	3.1148
270	5,506	767.9	28.11	1,185.77	2,790.1	1,604.3	2.9733	5.9244	2.9511
280	6,420	750.7	33.22	1,237.31	2,780.0	1,552.7	3.0646	5.8512	2.7866
290	7,445	732.3	39.18	1,290.52	2,766.2	1,475.7	3.1571	5.7754	2.6183
300	8,592	712.5	46.24	1,344.99	2,749.1	1,404.1	3.2513	5.6984	2.4471
310	9,870	690.6	54.64	1,401.97	2,726.9	1,324.9	3.3463	5.6155	2.2692
320	11,290	667.1	64.79	1,462.31	2,699.2	1,236.9	3.4447	5.5280	2.0833
330	12,865	640.2	77.20	1,526.00	2,664.8	1,138.8	3.5481	5.4343	1.8862
340	14,610	609.4	92.90	1,595.13	2,621.3	1,026.2	3.6561	5.3279	1.6718
350	16,540	572	113.6	1,671.39	892.5	892.5	3.7737	5.2045	1.4308
360	18,675	524	143.6	1,763.57	720.2	720.2	3.9152	5.0533	1.1381
370	21,050	448	200	1,985.14	448.4	448.4	4.1199	4.8164	0.6695

Critical temperature is 374.2 °C and critical pressure is 22,124 kPa.

Ice

Temperature T (°C)	Density ρ (kg/m³)	Thermal conductivity λ (W/m K)	Specific heat C_p (kJ/kg K)	Vapor pressure p (Pa)
0.01 (water)	999.8	0.552	4.221	611.7
0	916.2	2.22	2.050	611.2
−5	917.5	2.25	2.027	401.7
−10	918.9	2.30	2.000	259.9
−20	919.4	2.39	1.943	103.2
−30	920.0	2.50	1.882	38.0
−40	920.8	2.63	1.818	12.84
−50	921.6	2.76	1.751	3.94
−60	922.4	2.90	1.681	1.08
−70	923.3	3.05	1.609	0.261
−80	924.1	3.19	1.536	0.055
−90	924.9	3.34	1.463	0.0097
−100	925.7	3.48	1.389	0.0014

Air

- composition
- Prandtl
- thermal diffusion coefficient
- density
- dynamic viscosity
- thermal conductivity coefficient
- specific heat
- ratio of specific heats
- sound velocity

Composition of air

		Volume %	Weight %
Nitrogen	N_2	78.09	75.53
Oxygen	O_2	20.95	23.14
Argon	Ar	0.93	1.28
Carbon dioxide	CO_2	3×10^{-2}	5×10^{-2}
Neon	Ne	$1.8 \cdot 10^{-3}$	1.3×10^{-3}
Helium	He	5×10^{-4}	7×10^{-5}
Krypton	Kr	1×10^{-4}	3×10^{-4}
Xenon	Xe	8×10^{-6}	4×10^{-5}
Hydrogen	H_2	5×10^{-5}	3×10^{-6}
Ozone	O_3	1×10^{-6}	2×10^{-6}

The Prandtl number $Pr = \frac{\eta c_p}{\lambda}$ for air at atmospheric pressure.

T (°C)	Pr	T (°C)	Pr
0	0.715	250	0.68
20	0.713	300	0.68
40	0.711	350	0.68
60	0.709	400	0.68
80	0.708	450	0.685
100	0.703	500	0.69
120	0.700	600	0.69
140	0.695	700	0.70
160	0.690	800	0.715
180	0.690	900	0.725
200	0.685	1,000	0.735

Thermal diffusion coefficient $a = \frac{\lambda}{\rho c_p}$ for air at atmospheric pressure in m^2/s.

T (°C)	a	T (°C)	a
0	1.869×10^{-5}	250	6.028×10^{-5}
20	2.119×10^{-5}	300	7.028×10^{-5}
40	2.389×10^{-5}	350	8.111×10^{-5}
60	2.667×10^{-5}	400	9.194×10^{-5}
80	2.958×10^{-5}	450	1.031×10^{-4}
100	3.278×10^{-5}	500	1.142×10^{-4}
120	3.611×10^{-5}	600	1.381×10^{-4}
140	3.972×10^{-5}	700	1.619×10^{-4}
160	4.306×10^{-5}	800	1.858×10^{-4}
180	4.667×10^{-5}	900	2.100×10^{-4}
200	5.056×10^{-5}	1,000	2.350×10^{-4}

Density ρ of moisture-free air in kg/m^3.

T (°C)	ρ	T (°C)	ρ
0	1.293	180	0.779
10	1.247	200	0.746
15	1.226	250	0.675
20	1.205	300	0.616
25	1.185	350	0.566
30	1.165	400	0.524
40	1.127	450	0.488
60	1.060	500	0.456
80	1.000	600	0.404
100	0.946	700	0.363
120	0.898	800	0.329
140	0.854	900	0.305
160	0.815	1,000	0.277

Dynamic viscosity η of air in 10^{-5} Pa s at atmospheric pressure.

T (°C)	η	T (°C)	η
20	1.82	200	2.59
30	1.87	250	2.78
40	1.92	300	2.96
60	2.01	400	3.30
80	2.10	500	3.62
100	2.18	600	3.94
120	2.27	700	4.25
160	2.43		

Thermal conductivity coefficient λ of air in W/mK.

T (°C) P (Pa)	0	20	40	60	80	100
10^5	0.0243	0.0257	0.0271	0.0285	0.0299	0.0314
10^7	0.0298	0.0309	0.0320	0.0330	0.0342	0.0352
$2 \cdot 10^7$	0.0365	0.0370	0.0376	0.0381	0.0388	0.0395
$3 \cdot 10^7$	0.0442	0.0442	0.0443	0.0444	0.0445	0.0448
$4 \cdot 10^7$	0.0514	0.0511	0.0507	0.0502	0.0498	0.0493
$5 \cdot 10^7$	0.0587	0.0579	0.0571	0.0563	0.0555	0.0547

Specific heat C_p of air in kJ/kg K.

T (°C) P (Pa)	0	50	100	200	300	500	1000
10^5	1.006	1.009	1.013	1,027	1.048	1.094	1.194
$5 \cdot 10^6$	1.106	1.073	1.064	1.056	1.060	1.102	1.194
10^7	1.207	1.131	1.102	1.077	1.077	1.110	1.198
$1.5 \cdot 10^7$	1.282	1.182	1.135	1.098	1.094		
$2 \cdot 10^7$	1.349	1.223	1.165	1.119	1.110		
$2.5 \cdot 10^7$		1.253	1.194	1.140	1.127		
$3 \cdot 10^7$		1.270					

Ratio of specific heats C_p/C_v of air.

T (°C) P (Pa)	0	50	100	200	300	500	1,000
10^5	1.403	1.401	1.398	1.390	1.379	1.357	1.318
10^6	1.423	1.414	1.408	1.395	1.382	1.358	1.318
10^7	1.627	1.539	1.493	1.441	1.411	1.371	1.320

Sound velocity c in air in m/s.

P (Pa) T (°C)	10^4	5×10^4	10^5	2×10^5	5×10^5	10^6
17	341.39	341.46	341.49	341.60	341.92	342.55
27	347.22	347.26	347.33	347.46	347.82	348.49

Physical properties of some materials

Density, specific heat, and heat conductivity coefficient of some materials.

	Density ρ (10^3 kg/m³)	Specific heat C_p (kJ/kg K)	Heat conductivity coefficient λ (W/mK)
Asbestos, sheet	2.0	0.84	0.7
Asbestos, wool	0.58	0.84	0.2
Asphalt	1.1–1.5	0.92	0.74
Brick	1.4–1.6	0.92	0.4–0.5
Concrete	1.8–2.4	1.13	0.1–0.17
Cork	0.13	1.6–2.0	0.045
Glass	2.4–2.9	0.67–0.84	0.7–1.3
Granite	2.5–3.0	0.75	1.7–4.0
Graphite	1.9–2.3	0.84	5
Ice (0 °C)	0.92	–	2.09
Marble	2.6	0.88	2.8
Oakwood	0.7–0.9	2.4	0.19
Plaster	1.8	0.84	0.43
Quartz	2.6	0.80	1.1
Wood	0.5–0.9	1.9–2.7	0.1–0.2
Wool	0.11	1.4	0.036

Heat of combustion of some fuels (10^3 kJ/kg).

Acetylene	48.7	Fuel oil	40–42
Alcohol (95%)	27.0	Hydrogen	120
Ammonia	18.4	Methane	50
Brown coal	26–32	Paraffin oil	41–44
Butane	46.9	Petrol	42–45
Carbon monoxide	10.1	Petroleum	40–44
Coal	32–36	Propane	46.4
Ethane	47.5	Wood	19–20

Properties of natural gas.

Composition		
	CO_2	0.9%
	N_2	14.4%
	Methane	81.4%
	Ethane	2.7%
	Propane	0.4%
	Butane	0.2%

Heat of combustion	54.626 MJ/kg
Ignition temperature	670°C
Density	0.644 kg/m^3 (0 °C, 10^5 Pa)
Viscosity	1.10 × 10^{-5} Pa s (0 °C, 10^5 Pa)

Melting point and heat of melting of some materials.

	Melting point (°C)	Heat of melting (kJ/kg)
Aluminum	658	377
Antimony	630	163
Brass	900–980	168
Cadmium	321	46
Chromium	1,800	134
Cobalt	1,495	243
Copper	1,083	172
Glycerol	−20	176
Gold	1,064	67
Ice	0	335
Iron	1,200	205
Lead	327	23
Manganese	1,260	155
Mercury	−38.9	12
Nickel	1,452	235
Paraffin	52	147
Platinum	1,764	113
Silver	960	109
Sodium	63	59
Sulfur	113	38
Tin	232	59
Zinc	419	117

Properties of some liquids (20 °C).

		η (10⁻³ Pa s)	ρ (kg/m³)	c_p (kJ/kg K)	λ (W/mK)	a (10⁻⁹ m²/s)	Pr	Boiling point (°C)	γ (10⁻³ K⁻¹)	σ (10⁻³ N/m)
Acetone	C_3H_6O	0.422	791	2.16	0.162	94.8	5.63	56.2	1.43	23.7
Benzene	C_6H_6	0.649	879	1.73	0.140	92.1	8.02	80.1	1.06	28.9
Chloroform	CHC_{13}	0.667	1,490	0.993	0.145	98.0	4.57	61.7	1.28	27.1
Ethanol	C_2H_6O	1.201	789	2.40	0.167	88.0	17.3	78.5	1.10	22.8
Ether	$C_4H_{10}O$	0.243	714	2.34	0.138	82.6	4.12	34.5	1.62	17.0
Glycerol	$C_3H_8O_3$	1.499	1,260	2.37	0.285	95.4	12,465	290	0.50	63.4
n-Heptane	C_7H_{16}	0.413	684	2.20	0.140	93.0	6.49	98.4	1.24	—
n-Hexane	C_6H_{14}	0.308	660	2.23	0.137	93.1	5.01	68.95	1.35	18.4
Methanol	CH_4O	0.584	792	2.50	0.202	102.0	7.23	65.0	1.19	22.6
Nitric acid (98%)	HNO_3	0.890	1,512	1.09	0.267	162	3.63	83	1.24	42.7
n-Octane	C_8H_{18}	0.546	702	2.19	0.147	95.6	8.13	125.7	1.14	21.8
n-Pentane	C_5H_{12}	0.227	626	2.28	0.136	95.3	3.81	36.1	1.60	—
Propanol	C_3H_8O	2.231	804	2.35	0.158	83.6	33.2	97.4	0.98	23.8
Sulfuric acid	H_2SO_4	27.0	1,834	1.38	0.330	130	113	338	0.57	55.1
Toluene	C_7H_8	0.586	868	1.72	0.140	93.8	7.20	110.6	1.08	28.5
Water	H_2O	1.002	998	4.185	0.596	143	7.03	100	0.21	72.8

Properties of some gases at 25 °C and atmospheric pressure.

		η (10^{-6} Pa s)	ρ (kg/m³)	c_p (kJ/kg K)	λ (10^{-3} W/m K)	a (10^{-6} m²/s)	Pr –
Acetylene	C_2H_2	10.4	1.08	1.69	20	11	0.88
Air		18.3	1.185	1.007	26	22	0.71
Ammonia	NH_3	10.0	0.706	2.10	24	16	0.88
Carbon dioxide	CO_2	14.8	1.81	0.846	16	10	0.78
Chlorine	Cl_2	13.4	0.440	0.478	9	43	0.71
Helium	He	19.7	0.163	5.20	150	177	0.68
Hydrogen	H_2	8.90	0.082	14.3	190	162	0.67
Hydrogen sulfide	H_2S	12.6	1.41	1.01	14	9.8	0.91
Methane	CH_4	11.0	0.657	2.23	35	24	0.70
Nitrogen	N_2	17.7	1.15	1.04	26	21	0.71
Oxygen	O_2	20.5	1.31	0.913	27	23	0.69
Propane	C_3H_8	8.15	1.84	1.67	17	5.5	0.80
Sulfur dioxide	SO_2	12.8	2.68	0.608	9	26	0.86

Physical properties of some metals.

Metals (in alphabetical order)	Density ρ (10^3 kg/m³)	Melting point (°C)	Specific heat C_p (kJ/kg K)	Linear expansion coefficient a (10^{-6}/K)	Heat conductivity coefficient λ (W/mK)	Thermal conductivity a (10^{-3} m²/s)
Aluminum	2.70	660	0.922	24	237	0.095
Chromium	7.14	1,800	0.448	7	94	0.029
Cobalt	8.9	1,495	0.419	12	69	0.019
Copper	8.96	1,083	0.386	16.2	403	0.117
Gold	19.3	1,063	0.130	14.2	318	0.127
Iron	7.87	1,540	0.469	12	80.4	0.022
Lead	11.34	327	0.130	28.9	35.3	0.024
Magnesium	7.44	1,247	1.02	26	156	0.021
Manganese	7.2	1,244	0.478	22	8.3	0.0024
Mercury	13.55	– 39	0.138	182 (cub.)	8.3	0.0044
Nickel	8.85	1,455	0.461	13.2	90.9	0.022
Platinum	21.45	1,773	0.134	8.9	71.6	0.025
Plutonium	19.84	641	0.134	54	8	0.0030
Potassium	0.86	64	0.754	83	99	0.153
Silver	10.5	962	0.235	19.3	429	0.174
Sodium	0.97	98	1.23	70	1.34	0.0011
Tantalum	16.6	2,996	0.143	6.5	54	0.023
Tin	5.75	230	0.235	29	16.8	0.049
Titanium	4.5	1,660	0.524	8.5	0.2	0.00008
Uranium	19.05	1,132	0.117	13.4	0.25	0.0001
Wolfram	19.35	3,410	0.134	4.5	178	0.069
Zinc	7.12	419	0.390	30	116	0.042

Absorption coefficient *a* of various surfaces for heat radiation (in the order of increasing absorption coefficient).

Polished precious metals	0.02–0.05	Soft rubber	0.87
Polished copper	0.041	Oil paints	0.89–0.97
Polished brass	0.050	Oak wood	0.90
Polished aluminum	0.053	White enamel	0.90
Tin-coated iron	0.057	Lacquer paint	0.90
Nickel-coated iron	0.059	Gypsum	0.90
Brass sheet	0.069	Enamel paint	0.91
Aluminum sheet	0.071	Ice	0.91
Aluminum paint	0.3–0.4	Roofing paper	0.91
Ground sheet steel	0.66	Water	0.91
Iron plater, rusted	0.69	Red brick	0.94–0.95
cast iron	0.81	Glass	0.94
Coal	0.81	Hard rubber	0.95
Black lacquer	0.83	Asbestos	0.96
Polished marble	0.85	Blackbodies	1.00

Electrical conductivity *γ* of some aqueous solutions at 18 °C, *γ* in 10^{-6} S/m.

Solute	Concentration (wt%)	*γ*	Temperature coefficient at 18–26 °C $\frac{1}{\gamma_{18}}\left(\frac{\partial \gamma}{\partial T}\right)$
HCl	5	3,948	0.0158
	10	6,302	0.0156
	20	7,615	0.0154
	30	6,620	0.0152
	40	5,152	
H_2SO_4	5	2,085	0.0121
	10	3,915	0.0128
	20	6,527	0.0145
	30	7,388	0.0162
	40	6,800	0.0178
	50	5,405	0.0193
	60	3,726	0.0213
	70	2,157	0.0256
	80	1,105	0.0349
	90	1,075	0.0295
HNO_3	6.2	3,123	0.0147
	12.4	5,418	0.0142
	24.8	7,676	
	31.0	7,819	0.0139
	37.2	7,545	
	49.6	6,341	0.0157
	62.0	4,964	0.0157
NaOH	1	465	

(continued)

Solute	Concentration (wt%)	γ	Temperature coefficient at 18–26 °C	$\frac{1}{\gamma_{18}}\left(\frac{\partial \gamma}{\partial T}\right)$
	2	887		
	4	1,628		
	6	2,242		
	8	2,729		
	10	3,039		
	20	3,284		
	30	2,074		
	40	1,206		
	50	820		
NaCl	5	672	0.0217	
	10	1,211	0.0214	
	15	1,642	0.0212	
	20	1,957	0.0216	
	25	2,135	0.0227	

Solute	Concentration (wt%)	γ	Temperature coefficient at 18–26 °C
$NaNO_3$	5	436	0.0221
	10	782	0.0217
	20	1,303	0.0215
	30	1,606	0.0220
Na_2CO_3	5	451	0.0252
	10	705	0.0271
	15	836	0.0294
KCl	5	690	0.0201
	10	1,359	0.0188
	15	2,020	0.0179
	20	2,677	0.0168
K_2SO_4	5	458	0.0216
	10	860	0.0203

Coefficient of cubical expansion of some liquids at 20 °C $\gamma = \frac{1}{V}\left(\frac{\partial V}{\partial T}\right)$.

Liquid	$\gamma\ (°C^{-1})$
Acetone	1.487×10^{-3}
Alcohol	1.12×10^{-3}
Acetic acid	1.071×10^{-3}
Benzene	1.237×10^{-3}
Chloroform	1.273×10^{-3}
Methyl ether	1.656×10^{-3}
Phenol	1.090×10^{-3}
Glycerol	0.505×10^{-3}
Pentane	1.608×10^{-3}
Petroleum	0.955×10^{-3}
Turpentine	0.973×10^{-3}

Eutectic temperature of freezing mixtures with snow or ice.

Substances	wt%	Temperature (°C)
Copper sulfate	11.9	− 1.6
Potassium chloride	19.75	−10.5
Ammonium chloride	18.6	−15.8
Sodium chloride	22.4	−21.2
Magnesium chloride	21.6	−33.6
Nitric acid	32.7	−43
Calcium chloride	29.8	−55
Potassium hydroxide	31.5	−65
Chloric acid	24.8	−86

Properties of some plastics.

		Yield strength (10⁶ Pa)	Density (10³ kg/m³)	Linear thermal expansion coefficient (10⁻⁵ K⁻¹)	Specific heat (10³ J/kg K)	Thermal conductivity coefficient (W/mK)	Refractive index
CA	Cellulose acetate	42	1.3	8–9	1.3–1.7	0.26	1.45
PA	Polyamide 6/6 (nylon 6/6)	57	1.12–1.15	7–10	1.7–2.1	0.23–0.30	1.53
PC	Polycarbonate	62–67	1.20	6–7	1.2	2.0	1.587
LDPE	Low-density polyethylene	8–12	0.915–0.925	23	2.1–2.2	0.33–0.36	1.51
MDPE	Medium-density polyethylene	12–20	0.925–0.940	20	2.0–2.1	0.36–0.38	1.52
HDPE	High-density polyethylene	20–33	0.940–0.965	15–20	1.8–2.0	0.38–0.50	1.53
PMMA	Low-molecular-weight polymethylmethacrylate (perspex, plexiglass)	–	1.18	7–8	1.2–1.4	0.19	1.492
PMMA	High-molecular-weight polymethylmethacrylate (perspex, plexiglass)	–	1.18	7–8	1.2–1.4	0.19	1.492
PP	Polypropylene	30–33	0.905–0.907	16–18	1.7	0.22	1.50
PU	Polyurethane	–	1.21	19	1.7–2.1	0.36	–
PVC	Polyvinylchloride (rigid)	50–60	1.38–1.40	7–8	0.92–1.04	0.16	1.52–1.55
PVC	Polyvinylchloride (plasticized)	–	1.19	20	1.5	0.15	–
PTFE	Polytetrafluoroethylene (Teflon)	14–16	2.1–2.3	12–14	1.0	0.23–0.47	1.35

Crystallization temperature (°C)	Softening temperature acc. to Vicat (°C)	Maximum recommended service temperature (°C)	Transparency	chemical resistance										
				Strong acid	Weak acid	Strong alkali	Weak alkali	Alcohols	Esters	Ethers	Chlorinated hydrocarbons	Gasoline	Oil	
240–260	60	75	4	–	–	–	–	0	–	–	+/–	+	+	CA
250–255	250	90	3	–	–	0	+	0	+	+	0/–	+	+	PA
222–230	165	145	5	–	+	–	–	+/–	–	–	–	+	+	PC
108–115	–	80	3	+	+	+	+	+	–	0	–	0	+	LDPE
115–125	–	90	3	+	+	+	+	+	0	0	–	0/+	+	MDPE
125–135	60–70	95	3	+	+	+	+	+	+	0	–	+	+	HDPE
–	110	75	5	–	+	+	+	–	–	–	–	+	+	PMMA
–	115–120	100	5	–	+	+	+	–	–	–	–	+	+	PMMA
160–170	85–90	100	3	+	+	+	+	+	+/0	+	0	0	+	PP
170–175	160	80	2	–	0	+	+	+	+	+	0/–	+	+	PU
–	75–85	60	2	+	+	+	+	+	–	0	–	+	+	PVC
–	–	–	5	0	+	0	+	–	–	–	–	–	+	PVC
327	110	260	1	+	+	+	+	+	+	+	+	+	+	PTFE

Notation

Chemical resistance
+ resistant
0 somewhat resistant
– not resistant

Transparency
1 not transparent
2 opaque
3 semitransparent
4 transparent
5 clear

Refractive index of aqueous solutions at 20 °C.

wt%	Ethanol	Methanol	Acetone	Glycerol	HCl	NaCl	NaOH	Glucose
0.5	1.3333	1.3331	1.3334	1.3336	1.3341	1.3339	1.3334	1.3337
1	1.3336	1.3332	1.3337	1.3342	1.3353	1.3347	1.3358	1.3344
1.5	1.3339	1.3333	1.3341	–	1.3365	1.3356	1.3373	1.3351
2	1.3342	1.3334	1.3344	1.3353	1.3388	1.3365	1.3386	1.3358
2.5	1.3345	1.3335	1.3348	–	1.3399	1.3374	1.3400	1.3368
3	1.3348	1.3336	1.3352	1.3365	1.3411	1.3383	1.3414	1.3373
3.5	1.3351	1.3337	1.3355	–	1.3422	1.3391	1.3427	1.3380
4	1.3354	1.3339	1.3359	1.3376	1.3434	1.3400	1.3441	1.3387
4.5	1.3357	1.3340	1.3363	–	1.3445	1.3409	1.3454	1.3395
5	1.3360	1.3341	1.3366	1.3388	1.3457	1.3418	1.3467	1.3402
5.5	1.3364	1.3342	1.3370	–	1.3468	–	1.3481	1.3409
6	1.3367	1.3343	1.3373	1.3400	1.3480	1.3435	1.3494	1.3417
6.5	1.3370	1.3345	1.3377	–	1.3491	–	1.3507	1.3424
7	1.3374	1.3346	1.3381	1.3412	1.3503	1.3453	1.3520	1.3432
7.5	1.3377	1.3347	1.3384	–	1.3515	–	1.3533	1.3439
8	1.3381	1.3348	1.3388	1.3424	1.3526	1.3470	1.3546	1.3447
8.5	1.3384	1.3350	1.3392	–	1.3549	–	1.3559	1.3454
9	1.3388	1.3351	1.3395	1.3436	1.3561	1.3488	1.3572	1.3462
9.5	1.3392	1.3352	1.3399	–	1.3584	–	1.3585	1.3469
10	1.3395	1.3354	1.3402	1.3448	1.3561	1.3505	1.3597	1.3477
15	1.3432	1.3367	–	1.3509	1.3676	1.3594	1.3722	1.3555
20	1.3469	1.3381	–	1.3572	1.3792	1.3684	1.3840	1.3635
25	1.3505	1.3395	–	1.3637	1.3907	1.3776	1.3940	1.3719

Properties of foods.

Materials	Heat conductivity λ (W/mK)	Density ρ (kg/m³)	Viscosity η (m Pa · s)	Specific heat C_p (kJ/kg K)	Water content (wt%)
Apple sauce	0.69			4.0	–
Butter	0.20	998		2.3	15
Fish (fresh)	0.43			3.2	70
Fish (frozen)	1.22			1.7	70
Flour	0.45			1.8	8.8
Ice	2.3			2.0	100
Lamb	0.42			3.2	71
milk (whole)		1,030	2.12	3.9	87.5
Lamb milk (skimmed)	0.54	1,041	1.4	4.0	91
Oil – corn		921			
Oil – olive	0.17	919	84	2.0	
Oil – soybean		919	40		
Oranges	0.43			3.8	70
Pork	0.46			2.9	70
Potatoes	0.46			3.5	74

(continued)

Materials	Heat conductivity λ (W/mK)	Density ρ (kg/m^3)	Viscosity η (m Pa · s)	Specific heat C_p (kJ/kg K)	Water content (wt%)
Poultry (fresh)	0.50			3.3	74
Poultry (frozen)	1.68			1.6	74
Veal	0.49			3.2	70

Viscosities of foods Newton $\tau = -\eta\dot{\gamma}$

Materials		Temperature, T (K)	Viscosity, η (mPa s)
Milk	Whole	293	2.1
	Skimmed	293	1.4
Oil	Cotton seed	293	70
	Olive	293	84
	Peanut	323	24
	Soybean	303	40
	Honey	303	11,000

Power law $\tau = -K\dot{\gamma}^n$

Materials	Temperature, T (K)	Consistency, K (Ns^n/m^2)	Power law index, n (–)
Soups and sauces	286	3.5–5.6	0.51
Apple sauce	297	0.5	0.65
Tomato juice	305	18.7	0.40
Cream (48% fat)		0.127	0.65

Bingham $\tau - \tau_0 = -\eta\dot{\gamma}$

Materials	Temperature, T (K)	Viscosity, η (Pa s)	Yield stress, τ_0 (Pa)
Mayonnaise	303	0.63	85
Margarine	303	0.72	51
Mustard	303	0.25	38

Casson $\sqrt{\tau} - \sqrt{\tau_0} = \sqrt{|\dot{\gamma}|}$

Materials	Temperature, T (K)	Viscosity, η (Pa s)	Yield stress, τ_0 (Pa)
Tomato sauce		0.078	62.5
Chocolate (33.5% fat)			68
Chocolate (39% fat)			17

Compressibility of gases

Approximation according to Van der Waals:

$$Z \equiv \frac{Pv}{RT} = \frac{v}{v-b} - \frac{a}{RTv}$$

where

$$a = \frac{27R^2 T_{cr}^2}{64 P_{cr}}$$

$$b = \frac{RT_{cr}}{8P_{cr}}.$$

Approximation according to Redlich-Kwong:

$$Z \equiv \frac{Pv}{RT} = \frac{v}{v-b} - \frac{a}{RT^{3/2}(v+b)}$$

where

$$a = 0.42748 \times \frac{R^2 T_{cr}^{2.5}}{P_{cr}}$$

$$b = 0.08664 \times \frac{RT_{cr}}{P_{cr}}$$

Mixing rule for multicomponent systems:

$$a = \sum_{i=1}^{m} \sum_{j=1}^{m} y_i y_j \sqrt{a_i a_j}$$

$$b = \sum_{i=1}^{m} y_i b_i$$

Notation

Z: compressibility (–)

P: pressure (Pa)

V: molar volume (m^3/mol)

R: gas constant (8.314413 J/mol K)

T: temperature (K)

a: attraction parameter

b: "hard sphere" parameter

y: molar fraction in the vapor phase

P_{cr}: critical pressure (Pa)

T_{cr}: critical temperature (K)

V_{cr}: critical volume (m^3/mol)

Critical constants, boiling point, and molecular weight of various substances

Substances (in the order of increasing boiling point)	T_{cr} (K)	P_{cr} (10^5 Pa)	v_{cr} (m^3/mol)	Molecular weight (kg/kmol)	Atmospheric boiling point °C
Hydrogen	33.3	12.8	0.0650	2.02	−252.8
Nitrogen	126.1	33.5	0.0901	28.02	−195.8
Air	132.5	37.2	0.0934	28.96	−194.5
Carbon monoxide	134.2	35.0	0.0900	28.01	−191.5
Oxygen	154.3	49.7	0.0744	32.00	−183.0
Methane	191.1	45.8	0.0990	16.04	−161.5
Ethylene	283.1	50.5	0.124	28.05	−103.7
Ethane	305.6	48.3	0.148	30.07	−88.6
Acetylene	309.5	61.6	0.113	26.04	−81.5
Carbon dioxide	304.2	73.0	0.0957	44.01	−56.6
Hydrogen sulfide	373.6	88.9	0.0962	34.08	−60.4
Propylene	365.1	45.4	0.181	42.08	−47.7
Propane	369.9	42.0	0.200	44.10	−42.1
Ammonia	405.6	112.7	0.0729	17.03	−33.4
Isobutane	408.2	36.0	0.263	58.12	−11.7
Sulfur dioxide	430.6	77.8	0.1234	64.06	−10.0
n-Butane	425.2	37.5	0.255	50.12	−0.5
Isopentane	460.4	32.9	0.308	72.15	27.9
n-Pentane	469.6	33.3	0.311	72.15	36.1
Cyclopentane	511.8	44.5	0.260	70.13	49.3
n-Hexane	507.7	29.9	0.369	80.17	68.7
Methylcyclopentane	532.8	37.4	0.319	84.16	71.8
Benzene	562.6	48.6	0.263	78.11	80.1
Cyclohexane	553.0	40.2	0.311	84.16	80.7
n-Heptane	540.2	27.0	0.426	100.20	98.4
Water	647.1	218.3	0.0559	18.02	100.0
Methylcyclohexane	572.2	34.3	0.344	98.18	100.9
Toluene	593.9	41.6	0.328	92.13	110.6
n-Octane	568.6	24.6	0.486	114.22	125.7
Ethylbenzene	619.1	38.1	0.371	106.16	136.2
p-Xylene	618.2	35.0	0.371	106.16	138.4
m-Xylene	619.2	35.8	0.384	106.16	139.1
o-Xylene	632.2	36.9	0.384	106.16	144.4
n-Nonane	594.6	22.6	0.543	128.25	150.8
n-Decane	617.6	20.7	0.603	142.28	174.1
n-C_{15}	710	15.0	0.900	212.41	271
n-C_{20}	773	11.0	1.200	282.54	343
n-C_{30}	865	10.0	1.800	422.80	450
n-C_{40}	945	8.1	2.400	563.06	530
n-C_{50}	980	6.8	3.000	703.32	632

Antoine's parameters

Name	Formula	Parameters for Antoine's equation			Temperature range	ΔH_v	T_b
		A	B	C	°C	kJ/mol	°C
Acetone	C_3H_6O	14.3145	2756.22	228.060	−26 to 77	29.10	56.2
Acetic acid	$C_2H_4O_2$	15.0717	3580.80	224.650	24 to 142	23.70	117.9
Acetonitrile	C_2H_3N	14.8950	3413.10	250.523	−27 to 81	30.19	81.6
Benzene	C_6H_6	13.7819	2726.81	217.572	6 to 104	30.72	80.0
Isobutane	C_4H_{10}	13.8254	2181.79	248.870	−83 to 7	21.30	-11.9
n-Butane	C_4H_{10}	13.6608	2154.70	238.789	−73 to 19	22.44	-0.5
1-Butanol	$C_4H_{10}O$	15.3144	3212.43	182.739	37 to 138	43.29	117.6
2-Butanol	$C_4H_{10}O$	15.1989	3026.03	186.500	25 to 120	40.75	99.5
Isobutanol	$C_4H_{10}O$	14.6047	2740.95	166.670	30 to 128	41.82	107.8
tert-Butanol	$C_4H_{10}O$	14.8445	2658.29	177.650	10 to 101	39.07	82.3
Carbon tetrachloride	CCl_4	14.0572	2914.23	232.148	−14 to 101	29.82	76.6
Chlorobenzene	C_6H_5Cl	13.8635	3174.78	211.700	29 to 159	35.19	131.7
1-Chlorobutane	C_4H_9Cl	13.7965	2723.73	218.265	−17 to 79	30.39	78.5
Chloroform	$CHCl_3$	13.7324	2548.74	218.552	−23 to 84	29.24	61.1
Cyclohexane	C_6H_{12}	13.6568	2723.44	220.618	9 to 105	29.97	80.7
Cyclopentane	C_5H_{10}	13.9727	2653.90	234.510	−35 to 71	27.30	49.2
n-Decane	$C_{10}H_{22}$	13.9748	3442.76	193.858	65 to 203	38.75	174.1
Dichloromethane	CH_2Cl_2	13.9891	2463.93	223.240	−38 to 60	28.06	39.7
Diethyl ether	$C_4H_{10}O$	14.0735	2511.29	231.200	−43 to 55	26.52	34.4
1,4-Dioxane	$C_4H_8O_2$	15.0967	3579.78	240.337	20 to 105	34.16	101.3
n-Eicosane	$C_{20}H_{42}$	14.4575	4680.46	132.100	208 to 379	57.49	343.6
Ethanol	C_2H_6O	16.8958	3795.17	230.918	3 to 96	38.56	78.2
Ethylbenzene	C_8H_{10}	13.9726	3259.93	212.300	33 to 163	35.57	136.2
Ethylene glycol	$C_2H_6O_2$	15.7567	4187.46	178.650	100 to 222	50.73	197.3
n-Heptane	C_7H_{16}	13.8622	2910.26	216.432	4 to 123	31.77	98.4
n-Hexane	C_6H_{14}	13.8193	2696.04	224.317	−19 to 92	28.85	68.7
Methanol	CH_4O	16.5785	3638.27	239.500	−11 to 83	35.21	64.7
Methyl acetate	$C_3H_6O_2$	14.2456	2662.78	219.690	−23 to 78	30.32	56.9
Methyl ethyl ketone	C_4H_8O	14.1334	2838.24	218.690	−8 to 103	31.30	79.6
Nitromethane	CH_3NO_2	14.7513	3331.70	227.600	56 to 146	33.99	101.2
n-Nonane	C_9H_{20}	13.9854	3311.19	202.694	46 to 178	36.91	150.8
Iso-octane	C_8H_{18}	13.6703	2896.31	220.767	2 to 125	30.79	99.2
n-Octane	C_8H_{18}	13.9346	3123.13	209.635	26 to 152	34.41	125.6
n-Pentane	C_5H_{12}	13.7667	2451.88	232.014	−45 to 58	25.79	36.0
Phenol	C_6H_6O	14.4387	3507.80	175.400	80 to 208	46.18	181.8
1-Propanol	C_3H_8O	16.1154	3483.67	205.807	20 to 116	41.44	97.2
2-Propanol	C_3H_8O	16.6796	3640.20	219.610	8 to 100	39.85	82.2

ΔH_v	Vaporization enthalpy	kJ/mol
p^{sat}	Saturation pressure	kPa
T	Temperature	°C
T_b	Boiling temperature	°C

$$\ln p^{sat} = A - \frac{B}{T + C}$$

Based on Polling, Prauznitz, O'Connel: *Properties of Gasses and Liquids*, McGraw Hill New York 2001 and Gmehling, Oncken, Arit: *Vapor-Liquid Equilibrium Data*, DECHEMA Frankfurt/Main 1974-1990.

Solubility table

Notation
Example

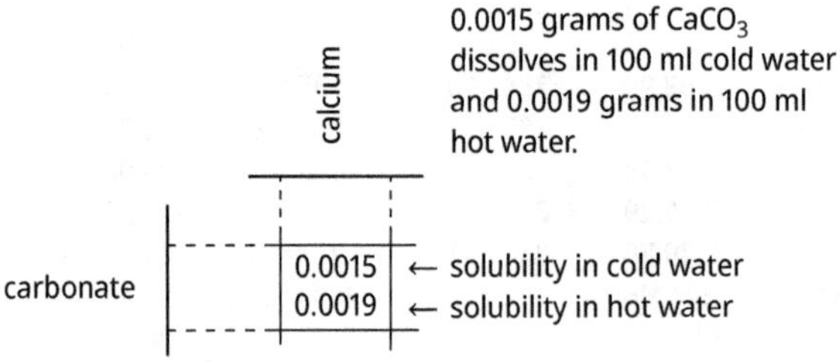

Fig. 5.1: Key to the solubility table.
a., dissolves in acids; d., decomposes in water; d.—i., decomposes in water (products are insoluble); i., insoluble in water; s., soluble in water (data unknown); v.s., very soluble in water; sl.s., slightly soluble in water; v.sl., very slightly soluble in water; s.d., soluble with some decomposition; *, chromium salts are found in two modifications of which the one dissolves better than the other; —, data unknown.

Solubility

	Sodium	Potassium	Ammonium	Magnesium	Calcium	Strontium	Barium	Aluminum	Chromium	Zinc	Manganese	Nickel
1. Acetate	119 170	253 >500	148 d.	36 >67	37 >67	s.	76.4 74	s. d.–i.	s. d.–i.	30 44.6	s.	16.6
2. Arsenate	26.7	v.s.	s. d.	i. 0.15	0.005	0.28 d.	0.06	a.	–	i.	i.	a.
3. Arsenite	v.s.	v.s.	v.s. d.	s. v.s.	a.	sl.s.	–	–	–	–	i.	a.
4. Borate	1.3 200	s.	s.	i. s.sl.	sl.s.	77	–	–	–	s.	–	–
5. Bromide	121	53 102	60 146	91 120	125 312	85 222	157 204	v.s.	• 200	471 675	300	113 155
6. Carbonate	7.1 45.5	112 156	100 d.	0.01	0.0015 0.0019	0.0011 0.065	0.002 0.006	–	–	a.	0.006	0.009
7. Chlorate	79 230	7.1 57	v.s.	57 74	178 v.s.	175 v.s.	27.4 111	v.s.	–	209	–	0.9
8. Chloride	35.7 39.8	34.7 57	30 76	53 73	60 159	44 101	31 59	70 s.d.	• i.(233)	432 615	151 656	64 87
9. Chromate	50 v.s.	63 79	40.5 d.	212 v.s.	16 20	0.12 3	0.003 0.004	–	–	a.	a.	a.
10. Cyanide	s. v.s.	v.s.	v.s. d.	s.	s. d.	v.s.	80	–	–	–	–	i.
11. Ferricyanide	18.9 67	33 78	v.s.	s.	v.s.	v.s.	s.	–	–	–	–	–
12. Ferrocyanide	32 161	28 91	s.	33	87 120	50 100	0.017 0.9	sl.s.	–	a.	a.	a.
13. Fluoride	4 v.s.	92 v.s.	v.s.	0.009 d.	0.0016 0.0017	0.012	0.17	s.	s.	s.	i. d.	0.02
14. Hydroxide	42 347	100 178	s.	0.0009 0.004	0.19 0.077	0.41 21.8	3.5 95	i.	i.	0.0004	a.	a.
15. Iodide	158.7 >257	128 208	154 250	100 165	182 426	165 383	200 269	s.d.	–	430 510	a.	124 188
16. Nitrate	73 180	32 247	118 871	200 v.s.	102 376	40 100	8.7 24.2	64 v.s.d.	s.	325	s.d.	240
17. Oxalate	3.7 6.33	33 –	2.5 11.8	0.07 0.08	0.0007 0.0014	0.005 5	0.007 0.022	i.–a.	s.	0.008	a.	a.
18. Oxide	d. v.s.	d. v.s.	–	0.0006 0.0008	0.12 d 0.07 d.	d.	d.	i.	i.	0.00016	i.	i.

(continued)

	Sodium	Potassium	Ammonium	Magnesium	Calcium	Strontium	Barium	Aluminum	Chromium	Zinc	Manganese	Nickel
19. Phosphate	v.s.	s.	v.s.	0.021	0.002	i.–a.	i.–a.	i.	a.	a.	a.	a.
20. Silicate	s.	s.	d.–i.	i.	0.0095	i.	i.	i.	–	i.	i.	i.
21. Sulfate	s.	6.8 / 24.1	71 / 194	26 / 74	0.24	0.011 / 0.011	0.0002 / 0.0004	31 / 98	• / i.–s.	86	52 / 70	29 / 83
22. Sulfide	s.	s. / v.s.	v.s. / d.	d.–i.	0.012 d / 0.461 d.	s.d.	d.	d.	d.	a.	a.	0.0003
23. Thiocyanate	v.s.	217	128 / v.s.	s.	v.s.	v.s.	43 / s.	–	–	s.	s.	s.

Table

	Cobalt	Iron-II	Iron-III	Silver	Lead	Mercury-I	Mercury-II	Copper	Bismuth	Cadmium	Tin-II	Tin-IV	Antimony	Gold	Platinum	
1.	s.	s.	d.–i.	1.0 / 2.52	45.6 / 200	0.75	25 / 100	7.2 / 20	d.–i.	v.s.	d.	–	–	–	–	1.
2.	a.	a.	a.	0.0008	a.	i.	v.sl.	a.	–	–	–	i.	–	–	–	2.
3.	a.	a.	a.	0.0012	a.	–	–	a.	–	–	–	–	–	–	–	3.
4.	–	–	–	sl.s.	a.	–	–	s.	–	a.	–	–	–	–	–	4.
5.	65 / 68	109 / 170	s.	8.4×10^{-6}	0.46 / 4,75	i.	0.5 / 25	v.s.	d.	57 / 162	85 / 222	s.d.	d.	s.	0.4	5.
6.	a.	0.0067	–	sl.s.	0.0001	–	–	a.	i.	i.	–	–	–	–	–	6.
7.	558	–	–	10 / 50	v.s.	s.	25	207 / v.s.	–	298 / 487	–	–	–	–	–	7.
8.	45 / 105	65 / 105	75 / 536	0.0001	0.67 / 3.34	0.002 / 0.001	3.6 / 61.3	71 / 108	d.	140 / 150	84 / 270	s.d.	600	68	v.s.	8.
9.	a.	–	–	0.0014	a.	v.sl.	sl.s.	i.	i.	a.	–	s.	–	–	–	9.
10.	i.	–	–	0.00002	s.l.s.	–	9.3 / 53	i.	–	1.7	–	–	–	d.	i.	10.
11.	s.	i.	–	0.00007	sl.s.	–	–	i.	–	a.	i.	i.	–	–	–	11.
12.	a.	–	i.	i.	i.	–	–	i.	–	a.	i.	i.	–	–	–	12.
13.	s.	sl.s.	sl.s.	182 / 208	0.06	s.d.	d.	sl.s.	–	4.35	v.s.	d.	384 / 563	–	s.	13.

(continued)

	Cobalt	Iron-II	Iron-III	Silver	Lead	Mercury-I	Mercury-II	Copper	Bismuth	Cadmium	Tin-II	Tin-IV	Antimony	Gold	Platinum	
14.	a.	a.	a.	i.	0.016	i.	i.	i.	i.							14.
15.	159 420	s.	–	2.6×10^{-6} 3×10^{-6}	0.0002 0.04 4.36	d. v.s.l.	– v.sl.	i. –	i. d.-i.	14. 80 127	1.32 3.55	d.	d.	s.	s.d.	15.
16.	134 v.s.	70 167	s.	122 952	38 126	d.	v.s.	v.s.	d.	109 326	d.	d.	–	–	–	16.
17.	a.	0.02 0.03	v.s.	0.0034	0.002	a.	a.	0.0025	a.	0.003 0.009	–	–	–	–	–	17.
18.	i.	i.	i.	0.0013 0.0053	0.002	i.	0.0052 0.040	i.	i.	a.	i.	i.	i.	i.	i.	18.
19.	a.	a.	a.	0.0006	a.	a.	a.	a.	a.	a.	i.	i.	i.	–	–	19.
20.	i.	i.	i.	–	i.	–	–	–	–	a.	–	–	–	–	–	20.
21.	36 83	16 50	sl.s. d.	0.57 1.41	0.003 0.006	0.06 0.09	d.	31 203	d.	75 60	18	v.s. d.	d.	s.s.	s.	21.
22.	0.0003	i.	i.	i.	i.	i.	i.	i.	i.	i.	i.	i.	i.	i.	i.	22.
23.	s.	s.	s.	0.00002	s.	i.	0.07	i.	–	a.	–	–	–	–	–	23.

Solubility parameters

- A mixture of two liquids approaches ideal thermodynamic behavior if the solubility parameters are close to each other.
- An amorphous polymer dissolves best in a solvent with the same solubility parameter.

Liquified gases at 90 K.

Substances	Solubility parameter $(10^3 \ (J/m^3)^{1/2})$
Nitrogen	1.08
Carbon monoxide	1.17
Argon	1.39
Oxygen	1.47
Methane	1.51
Carbon tetrafluoride	1.70
Ethane	1.94

Liquid solvents at 25 °C.

Substances	Solubility parameter 10^3 $(J/m^3)^{1/2}$
Perfluoro-*n*-heptane	1.23
Neopentane	1.27
Isopentane	1.39
n-Pentane	1.45
n-Hexane	1.49
I-Hexane	1.50
n-Octane	1.54
n-Hexadecane	1.64
Cyclohexane	1.68
Carbon tetrachloride	1.76
Ethyl benzene	1.80
Toluene	1.82
Benzene	1.88
Styrene	1.90
Tetrachloroethylene	1.90
Carbon disulfide	2.05
Bromide	2.35

Amorphous polymers near 25 °C.

Substances	Solubility parameter 10^3 $(J/m^3)^{1/2}$
Teflon	1.27
Polydimethyl silicone	1.49
Polyethylene	1.62
Polyisobutylene	1.66
Polybutadiene	1.76
Polystyrene	1.86
Polymethyl methacrylate	1.94
Polyvinyl chloride	1.99
Cellulose diacetate	2.23
Polyvinylidene chloride	2.50
Polyacrylonitrile	3.15

Diffusion coefficients

Diffusion coefficients of gases in water.

Gas		T (°C)	$\mathbb{D}$ $(10^{-9}$ m^2/s)
Acetylene	HC≡CH	17.5	1.69
Ammonia	NH_3	20	1.46
Argon	Ar	21.7	2.0
Carbon dioxide	CO_2	20	1.60
Chlorine	Cl_2	20	1.22
Ethylene	$H_2C=CH_2$	25.4	1.09
Helium	He	22.1	5.8
Hydrogen	H_2	21	3.81
Hydrogen sulfide	H_2S	25	1.36
Neon	Ne	22.1	2.8
Nitrogen	N_2	25	2.34
Nitrogen dioxide	NO_2	20	1.23
Nitrous oxide	N_2O	20	2.11
Oxygen	O_2	21	2.33
Propylene	$CH_3CH_2=CH_2$	25	0.90

Diffusion coefficients of gases and vapors in air at 25 °C and atmospheric pressure.

		$\mathbb{D}\ (10^{-6}\ m^2/s)$
Ammonia	NH_3	28
Benzene	C_6H_6	8.8
Carbon dioxide	CO_2	16.4
Ethanol	C_2H_6O	11.9
Ethyl ether	$C_4H_{10}O$	9.3
Hydrogen	H_2	41.0
Methanol	CH_4O	15.9
Oxygen	O_2	20.6
Toluene	C_7H_8	8.4
Water	H_2O	25.6
Xylene	C_8H_{10}	7.1

Diffusion coefficients of ions in water at very low concentrations at 25 °C.

		$\mathbb{D}\ (10^{-9}\ m^2/s)$
Bromide	Br^-	1.9
Cadmium	Cd^{2+}	0.7
Calcium	Ca^{2+}	0.8
Cesium	Cs^+	2.1
Chloride	Cl^-	2.0
Iodine	I^-	2.0
Sodium	Na^+	1.3
Sulfate	SO_4^{2-}	1.1

Henry's law

Henry's law constant for the solubility of gases in water

$P = Hx$

where P is the partial pressure of solute gas, x is the mole fraction of gas in water, and H is Henry's law constant.

Temperature (°C)	H, nitrogen (10^9 Pa)	H, oxygen (10^9 Pa)	H, air (10^9 Pa)	H, hydrogen (10^9 Pa)	H, CO_2 (10^7 Pa)	H, N_2O (10^8 Pa)	H, CH_4 (10^9 Pa)	H, H_2S (10^9 Pa)	H, NH_3 (10^5 Pa)	H, C_2H_2 (10^7 Pa)
0	5.36	2.58	4.38	5.87	7.62	1.00	2.27	2.71	2.08	7.47
5	6.05	2.95	4.95	6.16	9.16	1.24	2.62	3.19	2.24	88.65
10	6.77	3.31	5.56	6.44	10.9	1.48	3.01	33.72	2.41	9.86
15	7.48	3.69	6.15	6.70	12.8	1.76	3.41	4.28	2.59	11.2
20	8.15	4.06	6.73	6.92	14.8	2.09	3.81	4.89	2.80	12.5
25	8.77	4.44	7.30	7.16	17.1	2.43	4.18	5.52	3.02	13.9
30	9.36	4.81	7.81	7.38	19.5	2.80	4.45	6.17	–	15.2
35	9.98	5.14	8.34	7.52	–	–	4.92	6.85	–	–
40	10.5	5.42	8.82	7.61	24.5	3.60	5.27	7.55	–	18.3
45	11.0	5.71	9.23	7.70	–	–	5.58	8.25	–	–
50	11.5	5.96	9.59	7.75	29.8	4.51	5.85	8.95	–	21.4
60	12.2	6.37	10.2	7.75	35.7	–	6.34	10.43	–	24.3
70	12.7	6.72	10.6	7.71	42.0	–	6.75	12.05	–	–
80	12.8	6.96	10.8	7.65	–	–	6.91	13.68	–	–
90	12.8	7.08	10.9	7.61	–	–	7.01	14.59	–	–
100	12.8	7.10	10.8	7.55	–	–	7.10	15.0	–	–

Henry's law constant for the solubility of gases in ionic solutions

Ions influence the Henry's law constant according to:

$$\log \frac{H}{H_0} = -hI$$

where H_0 is Henry's law constant for pure water, I is the ionic strength $= \frac{1}{2}\sum \left(z_i^2 c_i\right)$, z_i is the charge of ion i, c_i is the concentration of ion i (mol/L), $h = h^+ + h^- + h_g$ according to the table.

Gas	h_g at:				
	0 °C	15 °C	25 °C	40 °C	60 °C
H_2		−0.008	−0.002		
O_2		0.034	0.022		
CO_2	−0.007	−0.010	−0.019	−0.026	−0.016
N_2O		0.003	0.000		
H_2S			−0.033		
NH_3			−0.054		
C_2H_2			−0.009		

Positive ions	h^+	Negative ions	h^-
H^+	0.000	OH^-	0.066
Na^+	0.091	Cl^-	0.021
K^+	0.074	NO_3^-	−0.001
NH_4^+	0.028	SO_4^{2-}	0.022
Mg^+	0.051	Br^-	0.012
Zn^{2+}	0.048	CO_3^{2-}	0.021
Ca^{2+}	0.053	I^-	0.005

Equilibrium data for hydrocarbons

Binary vapor-liquid equilibrium:

$$x_1 = \frac{1 - K_2}{K_1 - K_2} \qquad\qquad y_1 = K_1 x_1 = \frac{K_1(1 - K_2)}{K_1 - K_2}$$

Approximation for vapor pressure:

$$P_i^s = K_i P$$

Notation

x: molar fraction in the liquid phase

y: molar fraction in the vapor phase

K: equilibrium constant

P: pressure

P^s: saturation pressure

Equilibrium constants of hydrocarbons at 1 bar and 0–160 °C

mol% in vapor = K × mol% in liquid

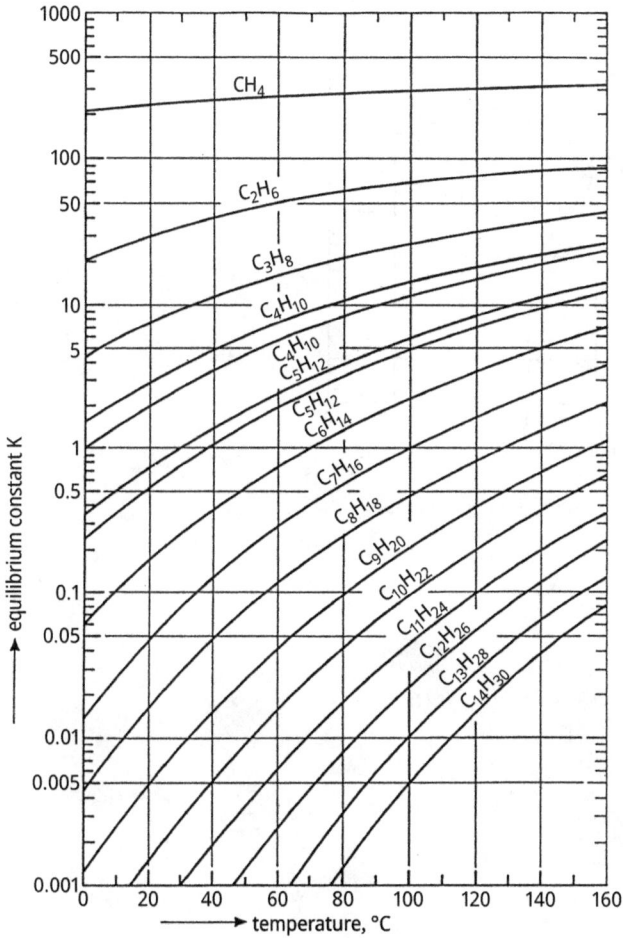

Fig. 5.2: Equilibrium constants at 1 bar and low temperatures.

Equilibrium constants of hydrocarbons at 1 bar and 140–300 °C

mol% in vapor = K × mol% in liquid

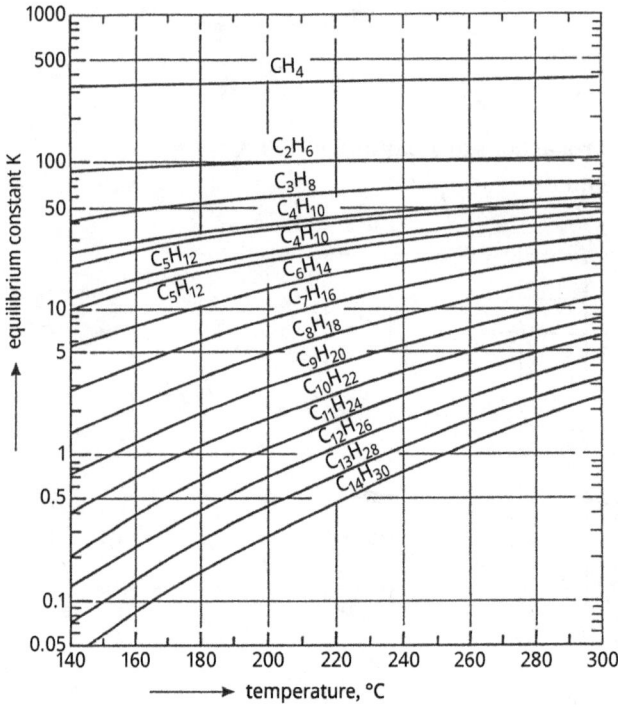

Fig. 5.3: Equilibrium constants at 1 bar and high temperatures.

Equilibrium constants of hydrocarbons at 10 bar and 0–160 °C

mol% in vapor = K × mol% in liquid

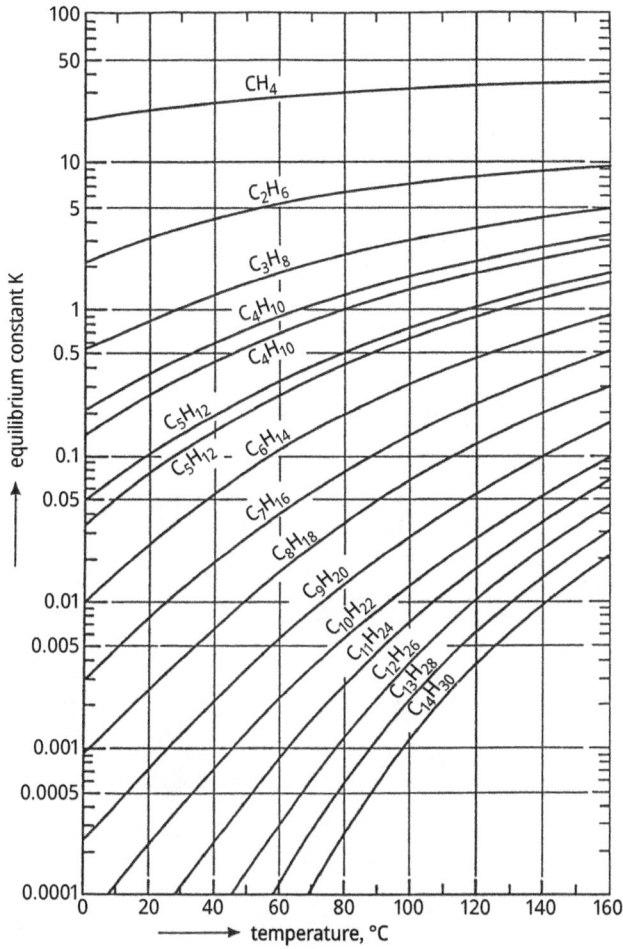

Fig. 5.4: Equilibrium constant at 10 bar and low temperature.

Equilibrium constants of hydrocarbons at 10 bar and 140–300 °C

mol% in vapor = K × mol% in liquid

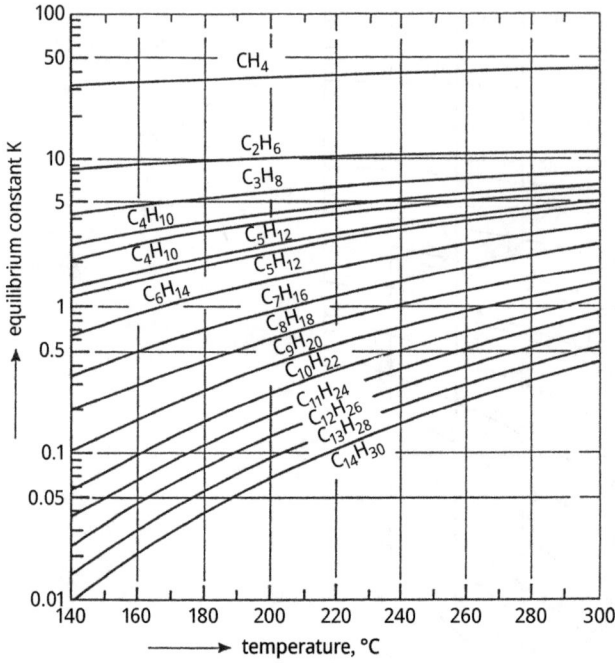

Fig. 5.5: Equilibrium constant at 10 bar and high temperature.

Dynamic viscosity

Dynamic viscosity η of ethanol-water solutions in mPa s

Wt% of ethanol	Temperature (°C)								
	0	10	20	30	40	50	60	70	80
10	3.31	2.18	1.54	1.16	0.91	0.73	0.61	0.51	0.43
20	5.32	3.17	2.18	1.55	1.16	0.91	0.74	0.61	0.51
30	6.94	4.05	2.71	1.87	1.37	1.05	0.83	0.68	0.57
40	7.14	4.39	2.91	2.02	1.48	1.13	0.89	0.73	0.60
50	6.58	4.18	2.87	2.02	1.50	1.16	0.91	0.74	0.61
60	5.75	3.77	2.67	1.93	1.45	1.13	0.90	0.73	0.60
70	4.76	3.27	2.37	1.77	1.34	1.06	0.86	0.70	0.59
80	3.69	2.71	2.01	1.53	1.20	0.97	0.79	0.65	0.57
90	2.73	2.10	1.61	1.28	1.04	0.85	0.70	0.59	0.50
100	1.78	1.46	1.19	1.00	0.83	0.70	0.59	0.51	0.44

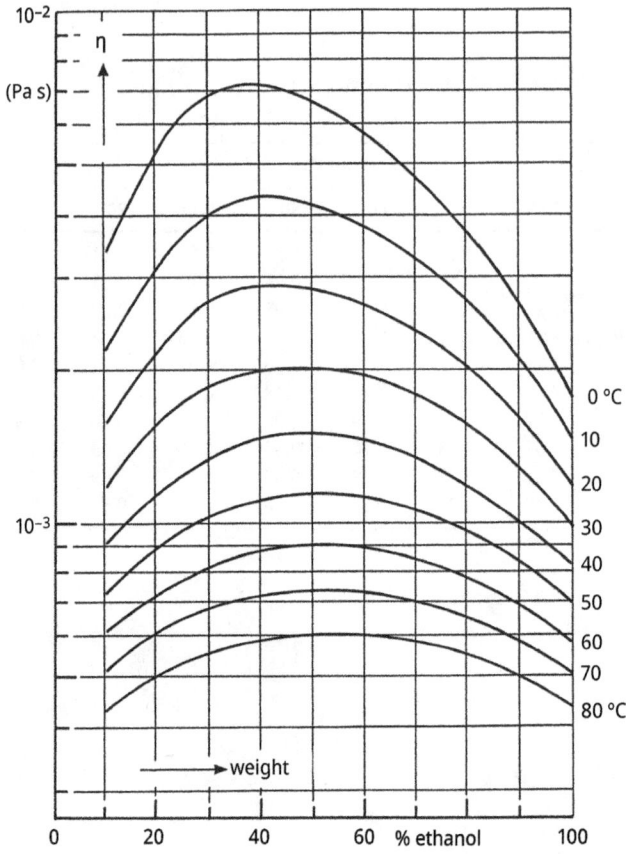

Fig. 5.6: Dynamic viscosities of ethanol - water mixtures (at different temperatures).

Dynamic viscosity μ of glycerol-water solutions in mPa s.

wt-% glycerol	20°C	25°C	30°C
5	1.143	1.010	0.900
10	1.311	1.153	1.024
15	1.517	1.331	1.174
20	1.769	1.542	1.360
25	2.095	1.810	1.590
30	2.501	2.157	1.876
35	3.040	2.600	2.249
40	3.750	3.181	2.731
45	4.715	3.967	3.380
50	6.050	5.041	4.247
55	7.997	6.582	5.494
60	10.96	8.823	7.312
65	15.54	12.36	10.02
70	22.94	17.96	14.32
75	36.46	27.73	21.68
80	62.0	45.86	34.92
85	112.9	81.5	60.05
90	234.6	163.6	115.3
91	278.4	184.3	134.4
92	328.4	221.8	156.5
93	387.7	262.9	182.9
94	457.7	308.7	212.0
95	545	366.0	248.8
96	661	435.0	296.7
97	805	522.9	354.0
98	974	629	424.0
99	1,197	775	511.0
100	1,499	945	624

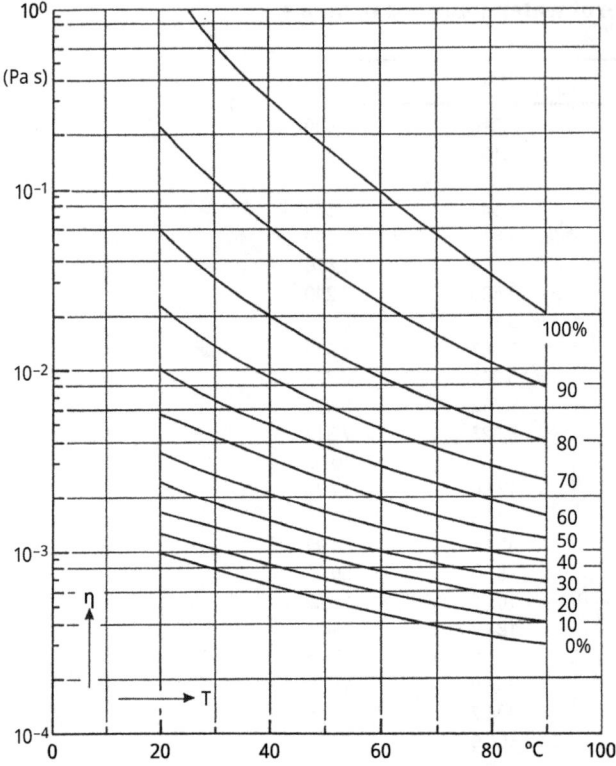

Fig. 5.7: Dynamic viscosities of glycerol - water mixtures (at different weight percentages of glycerol).

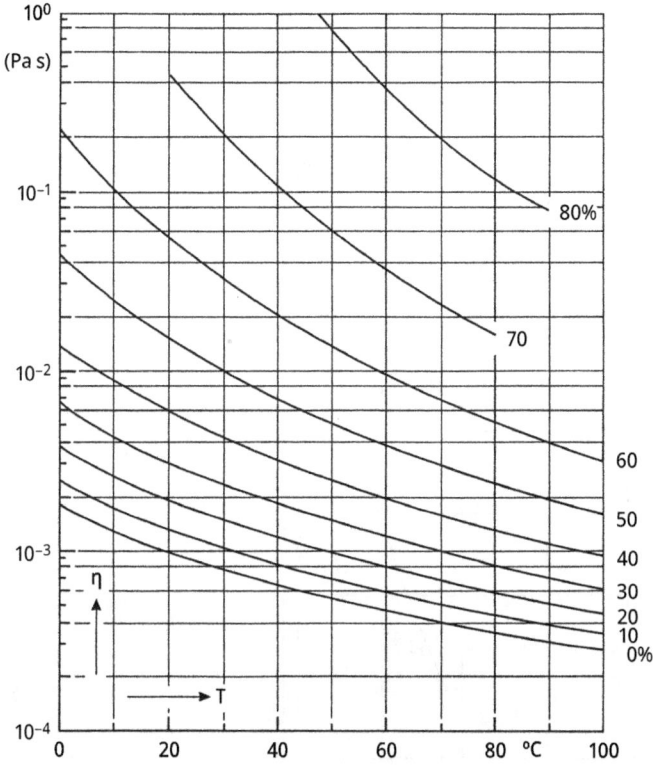

Fig. 5.8: Dynamic viscosities of corn syrup - water mixtures (at different weight percentages of cane sugar).

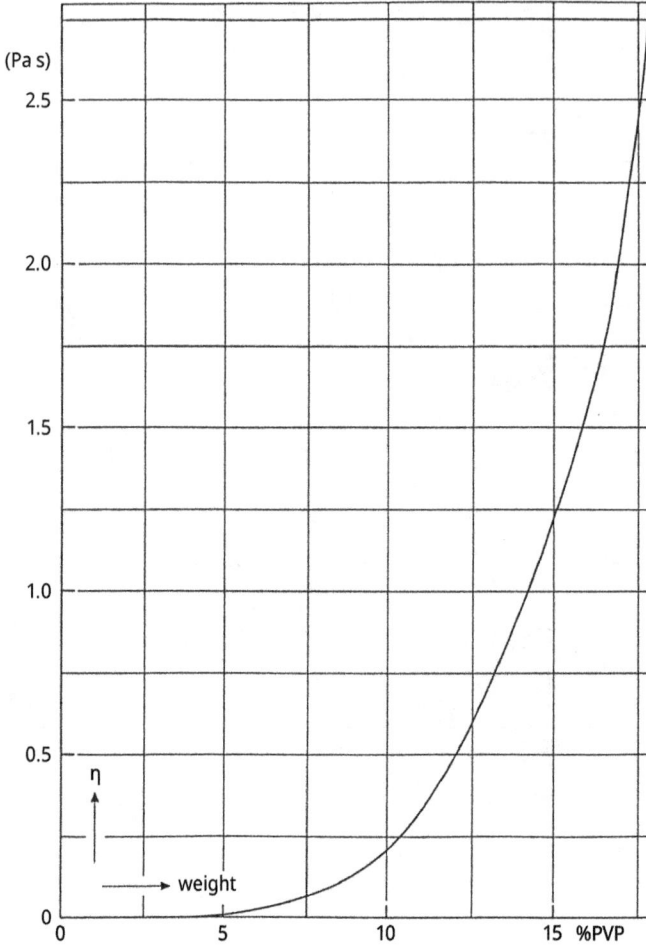

Fig. 5.9: Dynamic viscosity of polyvinylpyrrolidone (PVP) solution in tap water.

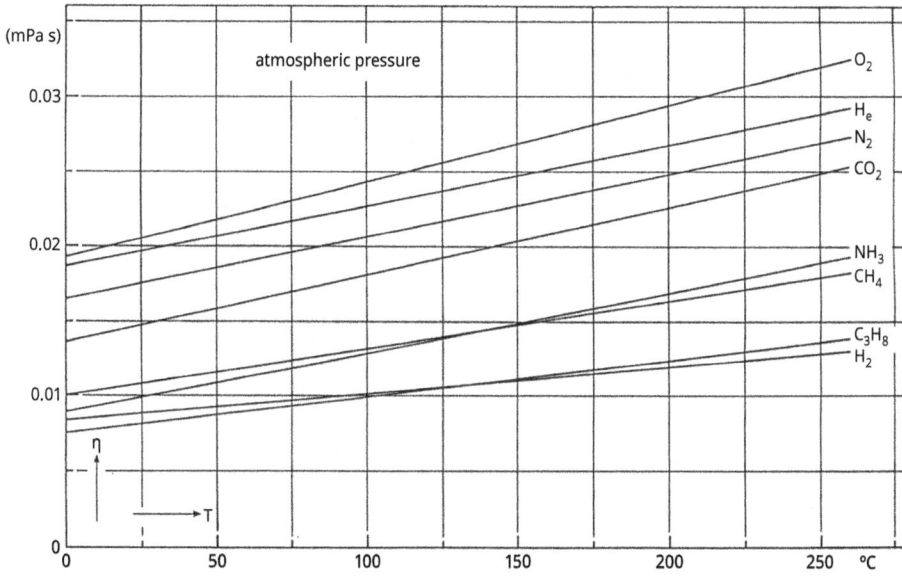

Fig. 5.10: Dynamic viscosity of gasses at atmospheric pressure.

Surface tension

Surface tension σ of some inorganic substances in water against air σ in 10^{-3} N/m.

Substances	Temperature (°C)	Concentration (wt%)					
		1	5	10	20	50	75
HCl	20	72.92	72.46	72.25	71.44		
H_2SO_4	18			74.1	75.2	77.3	72.86
HNO_3	20			72.65	71.1	65.43	
LiCl	18	73.2	74.75				
NaOH	20	73.17	74.6	77.3	85.8		
NaCl	18	72.7	73.95	75.51			
NaBr	18	72.95	73.6	74.4			
Na_2SO_4	18	72.75	73.82	75.15			
$NaNO_3$	30	71.4	72.1	72.8	74.4	79.8	
KCl	18	72.4	73.6	74.75	77.25		
KBr	18	72.85	73.25	73.9	75.25		
K_2SO_4	18	72.95	73.6				
KNO_2	20	72.95	73.6	74.35	76.0	81.7	94.6
KNO_3	18	72.7	73	73.6	75		
K_2CO_3	10	74.6	75.8	77.0	79.2	106.4	
NH_4OH	18	71.65	66.5	63.5	59.3		
NH_4Cl	18	72.6	73.3	74.5			
$MgCl_2$	18	72.75	73.8				
$MgSO_4$	18	73.1	73.75	74.25	77.6		
$CaCl_2$	18	72.72	73.7				
$BaCl_2$	30	71.2	74.35	74.5			
$Al_2(SO_4)_3$	18	73.6	74.5				
$MnSO_4$	15	73.31	73.65	74.85			
$FeSO_4$	18	73.6	73.62	73.9			
$CuSO_4$	30	71.25	71.8	72.37	73.5		
$AgNO_3$	18	73.5	73.9	74.4	75.25	78.9	
$ZnSO_4$	18	72.67	73.25	73.9	75.75		

Surface tension σ against air and interfacial tension σ_i against water of some organic liquids at 20 °C, σ and σ_i in 10^{-3} N/m.

Liquid		σ	σ_i
CCl_4	Carbon tetrachloride	26.66	45.0
CCl_3	Chloroform	27.13	32.80
CH_4O	Methanol	22.55	
C_2H_6O	Ethanol	22.27	
$n\text{-}C_3H_7O$	n-Propanol	23.8	
$n\text{-}C_4H_9O$	n-Butanol	24.6	1.58
$C_4H_{10}O$	Ethyl ether	17.10	10.70
$C_5H_{10}O$	Methyl propyl ketone	24.15	6.28
$n\text{-}C_5H_{11}O$	n-Pentanol	25.6	4.4
C_6H_5Cl	Phenyl chloride	33.08	37.41
$C_6H_5NO_2$	Nitrobenzene	43.38	25.66
C_6H_6	Benzene	28.86	35.0
C_6H_7N	Aniline	42.58	5.77
C_6H_{12}	Cyclohexane	26.54	50.3
$C_6H_{12}O$	Cyclohexanol	34.23	3.92
$C_6H_{12}O$	Ethyl propyl ketone	25.39	13.58
$C_6H_{12}O$	Methyl butyl ketone	25.49	9.73
C_6H_{14}	n-Hexane	18.43	51.10
C_7H_6O	Benzaldehyde	40.04	15.51
C_7H_8	Toluene	28.43	36.1
C_7H_8O	Benzyl alcohol	39.71	4.75
C_8H_8	Styrene	32.14	35.48
C_8H_{10}	Ethyl benzene	29.62	31.35
C_8H_{10}	o-Xylene	29.89	36.06
C_8H_{10}	p-Xylene	28.33	37.77
C_8H_{18}	n-Octane	21.77	50.81

Surface tension σ against air of some water-organic liquid solutions in 10^{-3} N/m.

Organic liquid		Temperature (°C)	Concentration (mol/L)							
			0.0078	0.0156	0.0312	0.0625	0.125	0.250	0.50	1.00
C_3H_6O	Acetone	15			69.8	68.5	66.6	63.6	59.4	54.1
C_6H_7O	Aniline	15				68.3	61.5			
$C_2H_4O_2$	Acetic acid	15				70.0	68.9	66.8	63.3	59.2
$C_4H_8O_2$	Butyric acid	15	69.8	68.6	65.8	59.8	55.1	47.9	40.1	32.4
$C_4H_{10}O$	Butanol	16	69.3	67.0	64.2	58.3	52.5	44.6	35.3	
$C_6H_{12}O_2$	Caproic acid	20	59.8	52.6	42.7					
$C_5H_{10}O$	Diethylketone	16	70.6	68.4	65.2	60.8	55.5	48.6		
C_2H_5O	Ethanol	18	69.4	68.1	65.2	59.4				
$C_5H_{12}O$	Isoamyl alcohol	18	67.4	62.2	55.1					
$C_4H_{10}O$	Isobutyl alcohol	18		69.8	66.6	60.9	54.2			
CH_4O	Methanol	18						70.2	68.4	65.1
C_4H_8O	Methyl ethyl ketone	19	71.0	70.3	68.5	65.5	62.0	57.3	50.3	43.0
C_6H_6O	Phenol	20	65.1	58.2	48.5	43.3				
$C_3H_6O_2$	Propionic acid	15		70.4	69.3	67.5	64.4	60.1	54.1	47.3
C_3H_8O	n-Propanol	15			68.9	66.8	63.3	57.7	50.5	42.4

Index

Abel's theorem 42
absolute humidity 18, 113–114
absorption coefficient 151
acetate 163
acetic acid 153, 184
acetone 149, 153, 156, 184
acetylene 147, 150, 160, 166
acid-base indicators 20
actinide series 10
activation energy 90, 92
adiabatic cooling line 113–114
adiabatic flow ideal gas 58
air 18, 144, 150, 160, 168
air-water system 113–114
alcohol 147, 153
alizarine, yellow 20
alphabet, Greek 7
aluminum 148, 150, 163
– sulfate 182
ammonia 147, 150, 160, 166–167
ammonium 163
– chloride 153, 182
– hydroxide 182
amyl alcohol 184
analogies between heat transfer and mass
        transfer 102
aniline 183–184
annular 121
Antoine's parameters 161
antimony 148, 164
apple sauce 156–157
aqueous solution 151, 156
Archimedes number 92
argon 144, 165–166
Arrhenius number 92
arsenate 163
arsenite 163
asbestos 128, 147, 151
asphalt 147
atmosphere, physical data 130
atomic mass 10
atomic number 10
average value 50
Avogadro's number 13
axial diffusion 92
azolithmine 20

balanced idealized materials 59
balances 54
barium 163
barium chloride 182
basic integrals 25
benzaldehyde 183
benzene 149, 153, 160, 166–167, 183
benzyl alcohol 183
Bernoulli, differential equations 33
Bernoulli's equations 2, 57
Bessel, differential equations 34
Bessel's functions 36
biaxial elongation 83
binary diffusion coefficient 18
binary vapor-liquid equilibrium 170
Bingham
– number 92
– plastic 84–86, 157
Biot number 2, 92
bismuth 164
black lacquer 151
blackbodies 151
Blake-Kozeny 104
Blasius flow 97
block 22
Bodenstein number 92
boiling point 160
Boltzmann's constant 13
Bond number 92
borate 163
brass 148, 151
brick 147, 151
brilliant cresyl blue 20
Brinkman number 92
bromide 163, 166–167
bromocresol blue 20
bromocresol purple 20
bromophenol blue 20
brown coal 147
bubble flow 123
bubble formation 124
bubble shape 118
bulk expansion coefficient 91
buoyancy force 94
Burke-Plummer 104
butane 147

https://doi.org/10.1515/9783111385341-006

butanol 184
butter 156
butyric acid 184

cadmium 148, 164, 167
calcium 163, 167
– chloride 153, 182
Candela 7
cane sugar 179
caproic acid 184
carbon dioxide 144, 150, 160, 166–167
carbon disulfide 166
carbon monoxide 147, 160, 165
– content of fuels 131
carbon tetrachloride 166, 183
carbon tetrafluoride 165
carbonate 163
Casson liquid 84, 158
Cauchy number 92
cellulose acetate 154
cellulose diacetate 166
Celsius 17
centrifugal force 93
cesium 167
characteristic length 90
characteristic process time 93
chemical reaction rate 93, 95
chemical resistance 155
Chilton-Colburn analogy 102
chlorate 163
chloric acid 151, 153
chloride 163, 167
chlorine 150, 166
chloroform 149, 153, 183
chlorophenol red 20
chocolate 158
chromium 148, 150, 162
classical Maxwell fluid 85
coal 147, 151
cobalt 148, 150, 164
codeformational derivative 49
color temperature 127
composition of air 144
composition of gas 147
compressibility 141
compressibility
– gas 158
concentration notation 53
concrete 147

conduction of heat 92
conductive heat transfer 94
conductive transport of heat 92
cone 23
consistency 84, 86, 157
content of fuels 131
continuity equation 57, 59, 63
contraction 107
convective heat transfer 94–95
convective mass transfer rate 92
convective transport of heat 93
conversion factor 15
copper 148, 150, 164
– sulfate 153, 182
cork 147
corn oil 156
corn syrup 179
corotational derivative 46
cotton seed oil 157
cream 157
cresol red 20
critical constant 160
critical pressure 159
critical temperature 159
critical volume 159
crystallization temperature 155
cube 22
cubic relations 21
curl of a vector field 48
curve fit 50
C_w coefficient 103
cyanide 163
cyclohexane 160, 166, 183
cyclohexanol 183
cyclopentane 160
cylinder 23
– segment 23

Damköhler number 92
Darcy's law 2
Dean number 93
Deborah number 93
deformation tensor 83
density 18, 53, 91, 139, 145, 147, 148, 150,
    154, 157
– water 142
derivatives 48, 49
– Jaumann 49
– Oldroyd 49

– scalar 48
– tensorial 49
derived S.I. units 7
dielectrical constant 138
diethylketone 184
differential equations 33
differentiation in tensor notation 47
differentiation in vector notation 47
diffusion coefficient 90, 166–167
diffusive mass transfer 95
diffusive mass transfer rate 95
dimensionless correlation 97
dimensionless numbers 92
dimensions 90
dimethyl yellow 20
Dirac pulse 43
dispersed flow 123
divergence of a vector field 47
drag coefficient 103, 105
drop shape 118
dynamic viscosity 91, 140, 145, 175, 177–181

earth, physical data 130
effective diffusion time 93
effective time for heat conduction 93
elastic force 93–94
elasticity number 93
electrical conductivity 138, 151
emission coefficient 128
enamel 151
energy balance 56, 61, 73
energy storage 133
enlargement 107
enthalpy of water 142
entropy of water 142
environmental data 130
Eötvös number 93
equation of continuity 54
equation of motion 55, 57
equilibrium data for hydrocarbons 170–174
erf x 40
Ergun 104
error function 39
ethane 147, 160, 165
ethanol 149, 156, 167, 175, 183–184
ethanol-water solution 175
ethyl benzene 183
ethyl ether 167, 183
ethyl propyl ketone 183

ethylbenzene 160
ethylene 160, 166
eutectic temperature 153
exit age distribution 87
expansion coefficient 137, 153
external heat transfer resistance 92
external mass transfer resistance 92
Eyring fluid 84

Fahrenheit 17
fall velocity drops in liquid 120
Fanning friction factor 93, 109
ferricyanide 163
ferrocyanide 163
ferrosulfate 182
Fick's equations 62
film flow 123
fish 156
flour 156
flow around obstacles 103
flow, dimensionless correlation 97
flow field tensor notation 83
flow resistance 103
fluidized bed drag coefficient 103
fluoride 163
food 156–157
Fourier 93, 115
– diagram 116, 117
– equations 61
– number 4, 6, 93, 115
free rising bubbles 118
freezing mixture 153
friction coefficient 107
friction factor 93, 109
frictionless flow 58
froth 121
Froude number 94
fuel 131, 147
fuel oil 147

Galileo number 94
gas constant 13, 90, 159
gas-liquid flow 121
glass 147, 151
glucose 156
glycerol 148–149, 153, 156
glycerol-water solution 177–178
gold 148, 150, 164
gradient of a scalar field 47

Graetz 99
– number 94
granite 147
Grashof
– number 94
gravitational acceleration 13, 90
gravitational force 92–93
Greek alphabet 7
greenhouse effect 131
gypsum 151

Hatta number 94, 125
heat capacity of fluid 95
heat conductivity 18, 157
– coefficient 147, 150
heat of combustion 148
heat of melting 148
heat of reaction 91, 93
heat of vaporization 137
– water 142
heat penetration 115
heat transfer 2, 95
– coefficient 3, 90
– dimensionless correlation 99
– number 102
Heaviside theorem 42
helium 144, 150, 166, 181
Henry's law 168
homogeneous, differential equations 34
honey 157
Hooke number 94
horizontal tubes 121
humidity diagram 113–114
hydraulic diameters 111
hydrocarbon 170–174
hydrodynamic boundary layer 95
– thickness 95
hydrogen 144, 147, 160, 166–168, 181
– chloride 156, 182
– nitrate 182
– sulfate 182
– sulfide 150, 160, 166
hydroxide 163

ice 143, 147–148, 151, 156
ideally mixed tank 87–88
ideally mixed tanks in series 89
ignition temperature 148
impellers 112

incompressible medium 58
inertia force 92–96
instationary heat transfer 4, 115
instationary mass transfer 6, 115
integrals 25
interfacial tension 183
internal age distribution 87
internal heat transfer resistance 92
internal mass transfer resistance
international system of units 7–9
invariants of a tensor 45
iodide 163, 167
iron 148, 150, 164
isobutane 160
isopentane 160, 166
isothermal flow ideal gas 58

Jaumann-Maxwell fluid 85
Jaumann's derivative 49
J-factor analogy 102

Kelvin 17
kinematic viscosity 91
kinetic energy 93, 95
Kirchhoff's law 126
krypton 144

lamb 156
laminar flow 86
– round tube 88
lanthanide series 10
Laplace transforms 41, 43
Laplacian of a field 48
lead 148, 150, 164
Lewis number 94
l-hexane 166
linear expansion coefficient 150
linear regression 50
linear thermal expansion 154
liquid solvent 166
liquids 149
liquified gas 165
lithium chloride 182

macrobalances 66
– Bernoulli equation 67
magnesium 150, 163
– chloride 153, 182
– sulfate 182

manganese 148, 150, 163
– sulfate 182
marble 147
margarine 157
mass balance 54, 59, 62, 77
mass fraction 53
mass penetration 115
mass transfer 102
– boundary layer thickness 94
– boundary layer 95
– coefficient 90
– dimensionless correlation 97
– with chemical reaction 94
– without chemical reaction 94
mathematical constants 13
Maxwell fluid 85
mayonnaise 157
melting point 148, 150
mercury 148, 150, 164
metals 150
methane 147, 150, 160, 165
methanol 149, 156, 167, 183–184
methyl butyl ketone 183
methyl ether 153
methyl ethyl ketone 184
methyl propyl ketone 183
methyl red 20
methylcyclohexane 160
methylcyclopentane 160
microscopic balances 54
– Bernoulli equation 57
– continuity equations 59
– Fick's equations 62
– for idealized materials 59
– Fourier equations 61
– momentum balances 59
– Navier-Stokes equations 59
milk 156–157
mixing rule 158
modified Bessel, differential equations 34
modified Bessel functions 38
modulus of elasticity 91
molar concentration 53
molar mass 18, 53
mole fraction 18, 53
molecular weight 160
momentum balance 59, 63, 68
Morton number 94, 118

multicomponent system 158
mustard 157
m-xylene 160

nabla identities 48
nabla operator 47
natural gas 147
Navier-Stokes equation 55, 59
n-butane 160
n-butanol 183
n-decane 160
neon 144, 166
neopentane 166
neutral red 20
Newtonian liquids 84–85, 118
n-heptane 149, 160
n-hexadecane 166
n-hexane 149, 160, 166, 183
nickel 148, 150, 163
nitramine 20
nitrate 163
nitric acid 149, 151, 153
nitrobenzene 183
nitrogen 144, 150, 160, 165–166,
        168, 181
– dioxide 166
nitrous oxide 166
n-nonane 160
n-octane 149, 160, 166, 183
nozzle 124
n-pentane 149, 160, 166
n-pentanol 183
number averaged molar mass 53
Nusselt number 94, 99, 102
nylon 6/6 154

oakwood 147
octahedron 22
oil 156–157
Oldroyd-Maxwell fluid 85
Oldroyd derivative 49
olive oil 156–157
orders of magnitude 1
oranges 156
organic liquid 183
organic solution 184
oxalate 163
oxide 163

oxygen 144, 150, 160, 165–168
o-xylene 160, 183
ozone 144

packed bed 104
packed bed, hydraulic parameters 111
paint 151
paraffin 148
– oil 147
peak sun hours (PSH) 132
peanut oil 157
Péclet number 87, 95
pentane 153
perfluoro-*n*-heptane 166
periodic table 10–12
perspex 154
petrol 147
petroleum 147, 153
pH indicators 19
phenol 153, 184
– red 20
phenolphthalein 20
phenyl chloride 183
phosphate 164
physical constants 13
physical properties 18, 147
– phases 94
pipe entrance 108
pipe exit 108
Planck's law 126
plaster 147
plastic 154
platinum 148, 150, 164
plexiglass 154
plug flow 88
plugs 121
plutonium 150
Poiseuille flow 97
polished marble 151
polyacrylonitrile 166
polyamide 6/6 154
polybutadiene 166
polycarbonate 154
polydimethyl silicone 166
polyethylene 154, 166
polyisobutylene 166
polymer 166
polymethylmethacrylate 154, 166

polypropylene 154
polystyrene 166
polytetrafluoroethylene 154
polyurethane 154
polyvinyl chloride 154, 166
polyvinylidene chloride 166
pork 156
potassium 150, 163
– bromide 182
– carbonate 182
– chloride 153, 182
– hydroxide 153
– nitrate 182
– nitrite 182
– sulfate 152, 182
potatoes 156
potential energy 95
poultry 157
Powell-Eyring fluid 84
power added 90, 95
power input 112
power law 84–86, 157
– index 84, 157
power number 95, 112
power transferred to kinetic energy 95
Prandtl 102
– fluid 84
– number 95, 136, 144, 149–150
Prandtl-Eyring fluid 84
precious metal 151
prefixes 9
pressure 18, 90
– drop in pipes 1
– characteristic throughput 86
process time 93
propane 147, 150, 160
propanol 149
properties of ice 143
properties of steam 142
properties of water 142
propionic acid 184
propylene 160, 166
PVP 180
p-xylene 160, 183
pyramid 22

quadratic relations 21
quartz 147

radiation 4, 151
Rankine 17
Rayleigh number 95
reaction constant 90
rectangular channel, hydraulic diameters 111
Redlich-Kwong 158
refractive index 141, 154, 156
relative humidity 18, 113–114
relative roughness 109
relaxation time 84, 91, 93
residence time distribution 87
Reynolds 107
– number 95, 102, 118
Reynolds number 111
rheogram 85
rheological model 84–85
Richardson number 95
rising velocity air bubbles in water 119
rising velocity drops in liquid 120
round tube, hydraulic diameters 111
rubber 151

salicyl yellow 20
sauce 157
scalar derivatives 48
Schmidt 102
– number 95
second invariant of the deformation
    tensor 84
second-order fluid 84
separation of variables 33
service temperature 155
shear stress 84
– energy wall 86, 93
Sherwood 102
– number 92, 95
shortage of elements 12
SI units 7–9
silicate 164
silver 148, 150, 164
– nitrate 182
simple elongation 83
simple shear 83
slug flow 123
slugs 121
sodium 148, 150, 163, 167
– bromide 182
– carbonate 152
– chloride 152–153, 156, 182

– hydroxide 151, 156, 182
– nitrate 152, 182
– sulfate 182
softening temperature 155
solar energy 132
solubility 162
– gas in ionic solution 169
– gas in water 168
– parameter 166
– table 162
sound velocity 146
soup 157
soybean oil 156–157
specific heat 17, 90, 140, 143, 146–147, 150,
    154, 157
specific mass 91
specific volume dry air 113
specific volume saturated air 113
speed of light 13
sphere 24
– sector 24
– segment 24
spiral diameter 90
spray 121
square tube, hydraulic diameters 111
standard deviation 50
standard pressure 13
standard temperature 13
Stanton number 95
steam 142
steel 151
Stefan-Boltzmann 12, 126
– constant 12
– law 126
stirrer 90, 112
– diameter 90
– power number 112
straight round tube 86
stratified 121
strontium 164
styrene 166, 183
substantial derivative 49
sulfate 153, 164, 167
sulfide 164
sulfur 148
– dioxide 150, 160
sulfuric acid 149, 151
surface calculations 22
surface tension 91, 121, 138, 182–184

– force 92–93, 95–96
Système Internationale 7–9

tantalum 150
Teflon 154
temperature 18
– conversion 17
– scale 17
tensor 45, 47, 49
– derviatives 49
– flow fields 82
– mathematics 44
– operation 46
tetrachloroethylene 166
tetrahedron 22
thermal boundary layer thickness 94
thermal conductivity 91, 143, 150, 154
thermal conductivity
– coefficient 139, 146
thermal diffusion 90
thermal diffusion
– coefficient 136, 145
thermal energy 92
– balances 61, 73
Thiele modulus 95
thiocyanate 164
thymol blue 20
thymolphthalein 20
tin 148, 150, 164
titanium 150
T-junction 108
toluene 149, 160, 166–167, 183
tomato juice 157
total derivative 49
total heat transfer 94
total mass transfer 95
transcendental functions integrals 30
transparency 155
triangular channel, hydraulic diameters 111
tropaeolin 00 20
truncated cone 23
truncated cylinder 23
truncated pyramid 22
tube 23
– flow 86
– system 107
turpentine 153
two-phase flow 121, 123

uniaxial elongation 83
unit step 43
uranium 150

valve 107
Van der Waals 158
vapor pressure 139
vapor pressure 143
vaporization entropy 142
veal 157
vector 44, 47
– mathematics 44
– operations 46
velocity of sound 142
viscoelastic force 96
viscosity 18, 84, 86, 157
viscous dissipation 92
viscous force 93–96
viscous stress 92
volume calculations 22
volume of an ideal gas 13

Waals, van der 158
water 18, 135, 142, 149, 151, 160, 167
– content 157
– vapor 19
wave 121
– flow 123
Weber number 96
Weissenberg number 96
Wien 13, 126
– displacement constant 13
– law 126
Williamson fluid 84
wolfram 150
wood 147, 151
wool 147

xenon 144
xylene 167

yield strength 154
yield stress 84, 86, 91–92, 157

zinc 148, 150, 163
zinc sulfate 182

α-naphthol benzene 20